Lecture Notes in Mathematics

A collection of informal reports and seminars
Edited by A. Dold, Heidelberg and B. Eckmann, Zürich

211

Théorie des Matroïdes

Rencontre Franco-Britannique
Actes 14 – 15 Mai 1970

Edité par C. P. Bruter
Faculté des Sciences, Brest/France

Springer-Verlag
Berlin · Heidelberg · New York 1971

AMS Subject Classifications (1970): 05 B 35

ISBN 3-540-05545-2 Springer-Verlag Berlin · Heidelberg · New York
ISBN 0-387-05545-2 Springer-Verlag New York · Heidelberg Berlin

Offsetdruck: Julius Beltz, Hemsbach/Bergstr.

W.T. TUTTE est né le 14 Mai 1917. Les Mémoires réunis dans ce volume, présentés à Brest les 14 et 15 Mai 1970 lors de la Rencontre Franco-Britannique sur la Théorie des Matroïdes, lui sont dédiés.

TABLE DES EXPOSES

*) L'exposé VIII ne figure pas dans ce recueil.

WHEELS AND WHIRLS

W.T. TUTTE

In graph theory there are several different kinds of "n-connection". Here n is a positive integer. The usual definition is that a graph is n-connected if it takes at least n vertices to separate two non-adjacent ones. A 1-connected graph is then simply a connected graph. Some embarrassment occurs in the case of a graph in which every pair of vertices are adjacent. Strictly we should say that such a graph is n-connected for every n, but this consequence of the definition is not acceptable to all graph theorists.

For loopless graphs we can go on to observe that a graph is 2-connected if and only if it is non-separable. If however we allow loops then this simple statement is non-longer justified. If a graph has a loop and at least one other edge we normally class it as separable, for it is the union of two proper subgraphs having at most a vertex in common. On the other hand loops are irrelevant in the above definition of n-connection. There are at least two methods of escape. We can, as they do in Michigan, restrict attention to graphs without loops. Or we can modify the definition of n-connection.

Consider the following variety of "n-connection". A graph is said to be k-separated if some k vertices separate k-edges from k other edges. A graph is n-connected if it is not k-separated for any k < n. With this definition we can still agree that the 1-connected graphs are the connected ones. But now it is also true that the 2-connected graphs are the non-separable ones. For a loop is separated by its end-vertex from any other edge. Hence a graph having a loop and at least one other edge is not 2-connected. Of course the loop-graph, consisting solely of a loop and its end, is 2-connected. Indeed it is n-connected for every n. This is acceptable, since the loop-graph is evidently non-separable.

Let us note in passing that, with both definitions, a graph that is n-connected is also k-connected, whenever $1 \leqslant k \leqslant n$.

Let us go on to the 3-connected graphs. With the second definition they must all be loopless, with the single exception of the loop-graph. On the first definition we can ignore loops together. On the first definition of n-connection it does not matter whether to join between two vertices is single or multiple. But with the second definition a graph that has a double join can have at most one other edge. The 2-circuit is 3-connected, and so is the θ-graph which consists of two vertices joined by three edges. We have seen also that the loop-graph is 3-connected. But on the second definition all other 3-connected graphs are "strict", that is have no loops or multiple joins.

For strict graphs it can be shown that the two definitions of 3-connection are equivalent. For general graphs the second definition seems more satisfactory to the author, and accordingly it will be used in what follows. Let us note however that the two definitions of n-connection do not agree, even for strict graphs, when $n > 3$. The second definition of n-connection has the advantage that, with it, n-connected planar graphs have n-connected duals. Another advantage is that it puts rather less emphasis on vertices. We may hope that this will make it easier to generalize to the theory of matroids.

We define the connectivity $\lambda(G)$ of a graph G as follows.

(i) If G is not connected, $\lambda(G) = 0$.

(ii) If there is a greatest positive integer n such that G is n-connected, then $\lambda(G) = n$.

(iii) In the remaining case $\lambda(G) = \infty$.

The graphs of infinite connectivity have been classified. They are the null graph, the vertex-graph (one vertex and no edges), the two connected graphs of one edge, the 2-circuit, the 3-circuit and the θ-graph.

There is a theory of 3-connected graphs based on the operations of deletion and contraction of edges. If A is an edge of a graph G we write G'_A for the graph obtained from G by deleting A, and G''_A for the graph derived from G by contracting A into a single vertex. Suppose G to be 3-connected. Then we say A is an essential edge of G if neither G'_A not G''_A is 3-connected. The basic problem is that of classification of all graphs G in which every edge is essential.

An example of such a graph is provided by any "wheel". A wheel of order n, where $n \geq 3$, is derived from an n-circuit C_n called the "rim" by adjoining a new vertex called the "hub" and then joining the hub to each vertex of the rim by a single edge called a "spoke". The deletion of any edge from a wheel leaves a vertex of valency 2. The two adjacent edges are separated from all other edges by two vertices. The contraction of any edge introduces multiple joins. The wheel itself can be shown to be 3-connected, but the above observations imply that each of its edges is essential. It has been shown (1) that the wheels are the only non-null 3-connected graphs for which all the edges are essential.

Let us consider the problem of generalizing this theorem to matroids. In matroid theory we apply the term "wheel" not to the graph described above but to the polygon-matroid of such a graph. Evidently the wheel thus defined is a self-dual matroid.

First we must generalize the notion of connectivity to matroids. To do this we must associate with each matroid M an integer $\lambda(M)$, and it seems best to require that $\lambda(M)$ and $\lambda(G)$ shall be equal when M is the polygon-matroid of G. Such an association was discovered by characterizing $\lambda(G)$ in terms of circuits and edges only. $\lambda(M)$ could then be similarly defined in terms if the circuits and cells of M.

Let M be a matroid on a set E. The definition of $\lambda(M)$ obtained by the above device runs as follows. For each pair $\{S, T\}$ of complementary subsets of E we write

$$\xi(M: S, T) = r(M) - r(M \times S) - r(M \times T) + 1.$$

Here $M \times S$ is the matroid on S defined by those circuits of M that are contained in S. The definition of $r(M)$ is as in (2). It is the minimum number of cells of M whose deletion destroys all the circuits of M.

We say that M is k-separated, where k is a positive integer, if there are complementary subsets S and T of E such that

$$\xi(M: S, T) = k,$$
$$\text{Min}(|S|, |T|) \geq k.$$

If there is a least positive integer k such that M is k-separated we call it the connectivity of M and denote it by $\lambda(M)$. If there is no such integer we write $\lambda(M) = \infty$. We say M is n-connected, where n is a positive integer, if $n \leq \lambda(M)$.

If M is the polygon-matroid of a graph G we can choose G to be connected. It is shown in (2) that then $\lambda(M) = \lambda(G)$. The proof is surprisingly complicated. However we may now claim that the notion of connectivity has been generalized from graphs to matroids. It can be shown that dual matroids have equal connectivities. We note in particular that the wheels are 3-connected.

If A is a cell of M we can define matroids M'_A and M''_A on the set $E - \{A\}$. The first is defined by those circuits of M that are contained in $E - \{A\}$, and the circuits of the second are the minimum non-null intersections with $E - \{A\}$ of the circuits of M. If M is the polygon-matroid of a graph G it can be verified that M'_A and M''_A are the polygon-matroids of G'_A and G''_A respectively. It seems appropriate therefore to define an essential cell of a 3-connected matroid M as a cell A such that neither M'_A nor M''_A

is 3-connected. In analogy with our graph-theoretical procedure we try to classify those 3-connected matroids M in which every edge is essential. Naturally the wheels are examples.

It is possible to construct an argument analogous to that used in graph theory to show that the (graphic) wheels are the only examples. But in the case of general matroids each step involves new complications. (See (2)). In particular the last step of the argument produces not only the wheels but another family of 3-connected matroids called the "whirls".

To obtain a whirl of order n, $n \geq 3$, we start with a wheel W of order n, the polygon-matroid of a wheel-graph G. The cells of the whirl are the cells of W, the edges of G. The circuits of W other than the one corresponding to the rim of G are recognized as circuits of the whirl. Any set of cells formed by adjoining one spoke of G to the set of edges of the rim is also recognized as a circuit of the whirl. The list of circuits of the whirl is now complete. It can be shown that this list satisfies the matroid axioms. It can also be shown that the whirl of order n is a self-dual matroid and that it is not the polygon-matroid of any graph. It can be verified that the whirls are 3-connected and that their cells are all essential.

The final result of the paper cited as Reference (2) is as follows. A non-null 3-connected matroid has all its cells essential if and only if it is a wheel or a whirl.

REFERENCES

[1] W. T. Tutte, A theory of 3-connected graphs. Nederl. Akad. Wetensch. Proc., 64 (1961), 441-455.

[2] W. T. Tutte, Connectivity in matroids, Can. J. Math., 18 (1966), 1301-1324.

GENERALIZED TRANSVERSAL THEORY

Richard A. Brualdi

Transversal Theory is that part of combinatorial theory that seeks criteria for the existence of transversals (or partial transversals) of families of sets where these transversals often are to satisfy certain additional properties. There is now quite an extensive literature on this subject, much of it coming in the last five or six years. As has been pointed out by Edmonds, transversal theory can be regarded as an instance of a more general theory which I shall call here Generalized Transversal Theory. This title reflects the attitude taken throughout the paper.

Let E be an arbitrary set, and let $\mathfrak{A}(I) = (A_i : i \in I)$ be a family of subsets of E, indexed by a set I. Recall that a transversal of $\mathfrak{A}(I)$ is a set $T \subseteq E$ for which there exists a bijection $\sigma : T \to I$ with $x \in A_{\sigma(x)}$ $(x \in T)$. A partial transversal is a transversal of a subfamily of $\mathfrak{A}(I)$. Each set A_i with $i \in I$ may be replaced by the collection $\rho_1(A_i)$ of subsets of A_i which consists of the empty set and the single element subsets of A_i. The collection $\rho_1(A_i)$ is then, trivially, the collection of independent sets of a matroid on E. More specifically, if A_i is a finite set, $\rho_1(A_i)$ is an instance of what I call a finite matroid; if A_i is an infinite set, it is an instance of what I call a rank-finite matroid (indeed, $\rho_1(A_i)$ has rank 1 if $A_i \neq \emptyset$ and rank 0 otherwise). A transversal of $\mathfrak{A}(I)$ can be regarded now as a set T for which there is a surjection $\sigma : T \to I$ with $\sigma^{-1}(i)$ a basis of the matroid $\rho_1(A_i)$ $(i \in I)$. If we assume that no set A_i is the empty set, it is no longer necessary to say σ is a surjection; we may say that T is a transversal of $\mathfrak{A}(I)$ provided there is a map $\sigma : T \to I$ with $\sigma^{-1}(i)$ a basis of $\rho_1(A_i)$ $(i \in I)$. A partial transversal of $\mathfrak{A}(I)$ is a set $F \subseteq E$ for which there exists a map $\rho : F \to I$ with

$\rho^{-1}(i)$ an independent set of the matroid $\rho_1(A_i)$ $(i \in I)$.

Thus we can say that a more general theory results when we replace a family of sets by a family of matroids and carry over the definitions of transversal and partial transversal as interpreted in matroid language above. Therefore, if $\mathfrak{E}(I) = (\mathfrak{E}^i : i \in I)$ is a family of matroids on a set E, we say that a set $T \subseteq E$ is a transversal of $\mathfrak{E}(I)$ provided there is a map $\sigma : T \to I$ with $\sigma^{-1}(i)$ a basis of $\mathfrak{E}^i$ $(i \in I)$ and we say that a set $F \subseteq E$ is a partial transversal of $\mathfrak{E}(I)$ provided there is a map $\rho : F \to I$ with $\rho^{-1}(i) \in \mathfrak{E}^i$ $(i \in I)$. From our experience in transversal theory it is suggested that in order to be able to prove existence theorems in generalized transversal theory, we shall have to assume that the matroids of the family are rank-finite matroids or even finite matroids. This is indeed so, and we shall present hardly any existential results about families of matroids that are permitted to have infinite rank. Indeed the basic problem of determining a criterion for a family of matroids of possibly infinite rank to have a transversal is, to my knowledge, unsolved - even for finite families. I shall, however, give necessary and sufficient conditions which apply under some restrictive circumstances. But when we restrict ourselves to families of finite matroids or rank-finite matroids, the situation is, happily, much different, and I would like to give a coherent account of much of what has been determined.

1. MATROIDS

Let me describe now those aspects of matroid theory which play a role in the subsequent discussion.

If E is an arbitrary set, a collection $\mathfrak{E}$ of subsets of E is a matroid on E provided (i) $\emptyset \in \mathfrak{E}$, (ii) $A \in \mathfrak{E}$, $A' \subseteq A$ imply $A' \in \mathfrak{E}$, and (iii) $A_1, A_2 \in \mathfrak{E}$, $|A_1| + 1 = |A_2| < \infty$ imply there is an $x \in A_2 \setminus A_1$ with $A_1 \cup x \in \mathfrak{E}$ †. The first

† For simplicity a set $\{x\}$ is denoted by x.

systematic study of matroids on finite sets was done by Whitney [32], while Tutte [30] has vigorously developed its theory. If a matroid ξ on E also satisfies (iv) $A \subseteq E$, $A' \in \xi$ for all finite $A' \subseteq A$ imply $A \in \xi$, then ξ is a matroid with finite character or, simply, a finite character matroid. A matroid, however, need not be a finite character matroid. The subsets of E which are members of ξ are called independent sets; all other subsets of E are dependent sets.

If ξ is a matroid on E, then a basis of ξ is a member of ξ which is maximal under set-theoretic inclusion (a maximal independent set). Bases need not exist as is seen by taking E to be an uncountable set and ξ to be the collection of all finite or countably infinite subsets of E. For a finite character matroid bases always exist (Zorn's lemma), and indeed every independent set can be enlarged to a basis. All bases of a finite character matroid have the same cardinal number [28]; indeed given two bases B_1, B_2 there is an injection $\tau : B_1 \to B_2$ with $\{B_2 \setminus \tau(x)\} \cup x$ a basis $(x \in B_1)$ [5]. It follows that if one basis of a finite character matroid is finite, all bases are and they all have the same cardinal number. We call such matroids rank-finite matroids. On the other hand, a matroid ξ on E is called a finite matroid provided $\{x \in E : \{x\} \in \xi\}$ is a finite set. Thus a finite matroid can be regarded as a matroid on a finite set.

If ξ is a matroid on E and $A \subseteq E$, then $\xi_A = \{F \subseteq A : F \in \xi\}$ is a matroid on A called the restriction of ξ to A. If ξ is a finite character matroid, so is ξ_A $(A \subseteq E)$. Since then all bases of ξ_A have the same cardinal number, there is a well-defined rank function r on the subsets of E. We set r(A) equal to the common cardinal number of the bases of ξ_A if this is a finite number, while we set $r(A) = \infty$ otherwise. The rank of ξ is r(E).

Besides restriction there are other ways of constructing new matroids from given ones. If ξ is a finite character matroid on E with $A \subseteq E$, let B be a basis of $\xi_{E \setminus A}$ and define

$$\mathcal{E}_{\otimes A} = \{F \subseteq A : B \cup F \in \mathcal{E}\}.$$

Then $\mathcal{E}_{\otimes A}$ is a matroid on E with finite character ([30], [6]). Moreover $\mathcal{E}_{\otimes A}$ is independent of the choice of basis B of $\mathcal{E}_{E \setminus A}$. We denote the rank function of $\mathcal{E}_{\otimes A}$ by r_A and call $r_A(A)$ the <u>contracted rank</u> of A. The matroid $\mathcal{E}_{\otimes A}$ is called the <u>contraction</u> of $\mathcal{E}$ to A.

If $\mathcal{E}$ is a matroid on E and k is a nonnegative integer then $(\mathcal{E})_k = \{A \subseteq E : A \in \mathcal{E},\ |A| \leq k\}$ is a finite-character matroid on E called the <u>truncation</u> of $\mathcal{E}$ at k [12]. If $\mathcal{E}_1$, $\mathcal{E}_2$ are two matroids on E, then $\mathcal{E}_1 \oplus \mathcal{E}_2 = \{A_1 \cup A_2 : A_1 \in \mathcal{E}_1,\ A_2 \in \mathcal{E}_2\}$ is a matroid on E called the <u>direct sum</u> of $\mathcal{E}_1$ and $\mathcal{E}_2$. If both $\mathcal{E}_1$ and $\mathcal{E}_2$ are finite character matroids, so is their direct sum.

In a finite character matroid $\mathcal{E}$ on E, a minimal dependent set is called a <u>circuit</u>. Because of the finite character property, circuits are finite sets. They are also nonempty sets with no one containing another properly. In addition, they satisfy the following <u>circuit property</u> ([30], [32]):

If C_1, C_2 are circuits with $a \in C_1 \setminus C_2$, $b \in C_1 \cap C_2$, there is a circuit C_3 with $a \in C_3 \subseteq \{C_1 \cup C_2\} \setminus b$.

Finally, if $\mathcal{E}$ is a matroid on E, an element $x \in E$ is a <u>loop</u> if $\{x\} \notin \mathcal{E}$ and a <u>coloop</u> if $A \in \mathcal{E}$ implies $A \cup x \in \mathcal{E}$. Thus loops can be part of no basis while coloops are in every basis. A matroid that has no coloops will be called <u>coloop-free</u>. If $\mathcal{E}$ is a finite character matroid on E and $A \subseteq E$ with B a basis of $\mathcal{E}_A$, then $A \cup \{x \in E : B \cup x \notin \mathcal{E}\}$ is the <u>span</u> of A in $\mathcal{E}$, denoted by sp(A). This definition is independent of the choice of basis of $\mathcal{E}_A$. Roughly speaking, the span of A consists of A along with those elements of $E \setminus A$ which depend on A.

Just a few words now about notation. If $(A_i : i \in I)$ is a family of sets with $K \subseteq I$, then $A(K) = \bigcup_{i \in K} A_i$. If a family $(X_j : j \in J)$ <u>partitions</u> a set X, we write

$X = \sum_{j \in J} X_j$. We use the notation $F \subset\subset E$ to signify that F is a finite subset of E. Finally if $(a_i : i \in I)$ is a family of nonnegative integers, then $\sum_{i \in I} a_i$ equals ∞ if $J = \{i \in I : a_i > 0\}$ is an infinite set and $\sum_{i \in J} a_i$ if J is a finite set.

2. FAMILIES OF MATROIDS

The starting point in transversal theory is furnished by the theorems of P. Hall and M. Hall Jr. which together give necessary and sufficient conditions for a family of sets, which satisfies some finitary conditions, to possess a transversal.

Theorem 2.1 Let $\mathfrak{A}(I) = (A_i : i \in I)$ be a family of subsets of a set E.

(i) (P. Hall [14] Assuming $|I| < \infty$, $\mathfrak{A}(I)$ has a transversal if and only if $|A(K)| \geq |K|$ $(K \subseteq I)$.

(ii) (M. Hall Jr. [15] Assuming $|I| = \infty$ and A_i is a finite set $(i \in I)$, $\mathfrak{A}(I)$ has a transversal if and only if $|A(K)| \geq |K|$ $(K \subset\subset I)$.

The result in generalized transversal theory which corresponds to P. Hall's theorem was obtained by Edmonds and Fulkerson ([12], [13]) while that corresponding to M. Hall Jr.'s theorem was obtained by me in [8] with the aid of the following theorems proved by means of the selection principle of Rado.

Theorem 2.2 Let $\mathcal{E}(I) = (\mathcal{E}^i : i \in I)$ be a family of finite matroids. Then $\mathcal{E}(I)$ has a transversal if and only if $\mathcal{E}(K) = (\mathcal{E}^i : i \in K)$ has a transversal for all $K \subset\subset I$.

Theorem 2.3 Let $\mathcal{E}(I) = (\mathcal{E}^i : i \in I)$ be a family of rank-finite matroids on a set E.

(i) (Edmonds and Fulkerson [13]) Assuming $|I| < \infty$, $\mathcal{E}(I)$ has a transversal if and only if

$$|F| \geq \sum_{i \in I} r_F^i(F) \qquad (F \subseteq E)$$

(ii) ([8]) Assuming $|I| = \infty$ and $\mathfrak{E}^i$ is a finite matroid ($i \in I$), $\mathfrak{E}(I)$ has a transversal if and only if

$$|F| \geq \sum_{i \in I} r^i_F(F) \qquad (F \subset\subset E) .$$

Actually in their theorems Edmonds and Fulkerson assume E is a finite set (and thus that the $\mathfrak{E}^i$ are finite matroids), but this is not necessary (see [8]).

A basic result in transversal theory is that if a family of sets has a transversal (no finiteness conditions imposed on the family), then every partial transversal can be enlarged to a transversal. For finite families this is implicit in a result of Hoffman and Kuhn ([16], [17]) and is also a consequence of a linking theorem of Mendelsohn and Dulmage [20]. For arbitrary families it is a consequence of the mapping theorem proved by Perfect and Pym [25]. The corresponding result in generalized transversal theory was proved by me in [8].

Theorem 2.4 *Let $\mathfrak{E}(I) = (\mathfrak{E}^i : i \in I)$ be a family of rank-finite matroids on a set* E. *Suppose $\mathfrak{E}(I)$ has a transversal. Then every partial transversal can be enlarged to a transversal. Indeed, if* B *is a transversal and* A *is a partial transversal, there exists a transversal* B^* *with* $A \subseteq B^* \subseteq A \cup B$.†

It was discovered by Edmonds and Fulkerson [13] that the collection $\underline{M}(\mathfrak{A}(I))$ of partial transversals of a family $\mathfrak{A}(I)$ of subsets of a set E is a matroid on E (see also [2]). This result was extended by Nash-Williams [24] and Edmonds to families of matroids.

Theorem 2.5 *Let $\mathfrak{E}(I) = (\mathfrak{E}^i : i \in I)$ be a family of matroids on a set* E. *Then the collection $\underline{M}(\mathfrak{E}(I))$ of partial transversals of $\mathfrak{E}(I)$ is a matroid on* E.

---

† In [8] we assumed the $\mathfrak{E}^i$ were finite matroids but the same proof works for rank-finite matroids.

(i) If $\{i \in I : \{e\} \in \mathfrak{E}^i\}$ is a finite set $(e \in E)$ and each $\mathfrak{E}^i$ is a finite character matroid $(i \in I)$, then $M(\mathfrak{E}(I))$ is a finite character matroid.

(ii) If r^i is the rank function of $\mathfrak{E}^i$ $(i \in I)$ and r is the rank function of $M(\mathfrak{E}(I))$, then

$$r(A) = \min\{\sum_{i \in I} r^i(F) + |A \setminus F| : F \subseteq A\}$$

for all $A \subseteq E$.

Of course in (ii) we need only take the minimum over the cofinite subsets F of A, that is, those subsets F of A with $|A \setminus F|$ finite. The finite character property of (i) follows easily from Rado's selection principle (see [26] or [8]). Part (ii) for a finite collection of matroids on a finite set is contained in Nash-Williams' work [24] and is also an immediate consequence of the matroid partition theorem of Edmonds and Fulkerson [13]. Part (ii) for an arbitrary family of matroids and arbitrary A is proved by Pym and Perfect [26] and by me [8]. The methods used are quite different.

In a finite character matroid all bases have the same cardinal number. This is not necessarily true in arbitrary matroids. In fact, Dlab [11] starting with a different set of axioms has shown that cardinal numbers of bases can be chosen arbitrarily. If $\mathfrak{E}(I) = (\mathfrak{E}^i : i \in I)$ is a family of finite character matroids, the matroid $M(\mathfrak{E}(I))$ need not have finite character and thus need not have even one basis. However, bases, if they exist, must have the same cardinal number.

Theorem 2.6 Let $\mathfrak{E}(I) = (\mathfrak{E}^i : i \in I)$ be a family of finite character matroids and suppose there are bases B_1 and B_2 of $M(\mathfrak{E}(I))$. Then there is an injection $\sigma : B_1 \to B_2$ such that $\{B_2 \setminus \sigma(x)\} \cup x$ is a basis $(x \in B_1)$. In particular, B_1 and B_2 have the same cardinal number.

We prove here a stronger statement. Let $A, B \in \underline{M}(\mathfrak{E}(I))$ with B a basis. Thus $A = \sum_{i \in I} A_i$ with $A_i \in \mathfrak{E}^i$ $(i \in I)$ and $B = \sum_{i \in I} B_i$ with $B_i \in \mathfrak{E}^i$ $(i \in I)$. Consider the family of matroids $\tilde{\mathfrak{E}}(I) = (\tilde{\mathfrak{E}}^i : i \in I)$ where $\tilde{\mathfrak{E}}^i = \mathfrak{E}^i_{A_i \cup B_i}$ $(i \in I)$. Obviously $A, B \in \underline{M}(\tilde{\mathfrak{E}}(I))$ with B a basis, and we may consider $\underline{M}(\tilde{\mathfrak{E}}(I))$ as a matroid on $A \cup B$. Let $x \in A \cup B$ be such that $\{x\} \in \tilde{\mathfrak{E}}^i$ for some $i \in I$; then $x \in A_i$ or $x \in B_i$. But since $(A_i : i \in I)$ partitions A and $(B_i : i \in I)$ partitions B, x is an element of at most one of the A_i's and at most one of the B_i's. Hence

$$|\{i \in I : \{x\} \in \tilde{\mathfrak{E}}^i\}| = 1 \text{ or } 2 \qquad (x \in A \cup B).$$

By Theorem 2.5(i), $\underline{M}(\tilde{\mathfrak{E}}(I))$ is a finite character matroid so that, as easily follows from the result in [5], there is an injection $\sigma : A \to B$ with $\{B \setminus \sigma(x)\} \cup x$ a basis of $\underline{M}(\tilde{\mathfrak{E}}(I))$ $(x \in A)$. If $B' = \{B \setminus \sigma(x)\} \cup x$ were not a basis of $\underline{M}(\mathfrak{E}(I))$, then there would exist $y \notin B'$, $y \neq \sigma(x)$ such that $B' \cup y \in \underline{M}(\mathfrak{E}(I))$. But now there is an injection $\tau : B' \cup y \to B$. Since $|B \setminus \{B' \cup y\}| = 1$, $|\{B' \cup y\} \setminus B| = 2$, this is impossible. Thus B' is a basis of $\underline{M}(\mathfrak{E}(I))$.

Besides the matroid $M(\mathfrak{E}(I))$ arising from a family $\mathfrak{E}(I)$ of matroids on a set E, there is often another matroid present which is something like a dual and indeed coincides with the dual if E is a finite set. The result for families of sets was proved by Scrimger and me [5]. Let $\mathfrak{A}(I) = (A_i : i \in I)$ be a family of subsets of a set E. Call a subset J of I <u>maximal</u> provided $\mathfrak{A}(J) = (A_i : i \in J)$ has a transversal but $\mathfrak{A}(K)$ does not have a transversal for all sets K with $J \subseteq K \subseteq I$, $J \neq K$. If $\underline{M}^*(\mathfrak{A}(I))$ is the collection of subsets F of E for which there is a maximal J with $\mathfrak{A}(J)$ having a transversal $T \subseteq E \setminus F$, then $\underline{M}^*(\mathfrak{A}(I))$ is a matroid on E. If $\mathfrak{A}(I)$ has a transversal, then the matroid $\underline{M}^*(\mathfrak{A}(I))$ consists of those subsets of E whose complement (in E) contains a transversal of $\mathfrak{A}(I)$. (This is not the same, in general, as the collection of subsets of E which are contained in the complement of a basis of $\underline{M}(\mathfrak{A}(I))$; indeed $\underline{M}(\mathfrak{A}(I))$ need not have any bases for $\underline{M}^*(\mathfrak{A}(I))$

to be defined.) What was not noted in [9], but nevertheless is true, is that if $\mathfrak{A}(I)$ is a family of finite sets with a transversal, then $\underline{M}^*(\mathfrak{A}(I))$ is a finite character matroid. From this it follows that every transversal of $\mathfrak{A}(I)$ contains a minimal (in the set-theoretic sense) transversal of $\mathfrak{A}(I)$. This is a result complementary to the result that $\underline{M}(\mathfrak{A}(I))$ is a finite character matroid if each element of E is a member of only finitely many sets of the family $\mathfrak{A}(I)$.

Theorem 2.7 Let $\mathfrak{E}(I) = (\mathfrak{E}^i : i \in I)$ be a family of finite character matroids on E which has a transversal. Let $\underline{M}^*(\mathfrak{E}(I))$ consist of all those subsets F of E for which $E \setminus F$ contains a transversal of $\mathfrak{E}(I)$. Then $\underline{M}^*(\mathfrak{E}(I))$ is a matroid on E. If $\mathfrak{E}^i$ is a finite matroid $(i \in I)$, then $\underline{M}^*(\mathfrak{E}(I))$ is a finite character matroid.

The only property for matroids that is not obviously satisfied by $\underline{M}^*(\mathfrak{E}(I))$ is the replacement property (iii) in the definition of a matroid. Thus let $A_1, A_2 \in \underline{M}^*(\mathfrak{E}(I))$ with $|A_1| + 1 = |A_2| < \infty$. Hence there exists a transversal B_j of $\mathfrak{E}(I)$ with $A_j \subseteq E \setminus B_j$ and $B_j = \sum_{i \in I} B_j^i$ $(j = 1, 2)$ where B_j^i is a basis of $\mathfrak{E}^i$ $(i \in I,\ j = 1, 2)$. If there were an $x \in \{A_2 \setminus A_1\} \setminus B_1$, then $A_1 \cup x \subseteq E \setminus B_1$ so that $A_1 \cup x \in \underline{M}^*(\mathfrak{E}(I))$. Thus we may assume that $A_2 \setminus A_1 \subseteq B_1$. Let $D_1 = A_2 \cap B_1$, $D_2 = A_1 \cap B_2$ so that it now follows that $|D_1| > |D_2|$. Now

$$D_1 = \sum_{i \in I} D_1 \cap B_1^i = \sum_{i \in I} D_1^i \qquad (D_1^i \in \mathfrak{E}^i,\ i \in I)$$

$$D_2 = \sum_{i \in I} D_2 \cap B_2^i = \sum_{i \in I} D_2^i \qquad (D_2^i \in \mathfrak{E}^i,\ i \in I).$$

Since $|D_1| > |D_2|$, there exists a $k \in I$ with $|D_1^k| > |D_2^k|$. Extend $B_1^k \setminus D_1^k$ to a basis of $\mathfrak{E}^k$ by means of elements in $B_2^k \setminus B_1^k$; thus there exists $C \subseteq B_2^k \setminus B_1^k$ such that $\{B_1^k \setminus D_1^k\} \cup C = \widetilde{B}^k$ is a basis of $\mathfrak{E}^k$. Then $\widetilde{B}^k \setminus B_1^k = C$ while $B_1^k \setminus \widetilde{B}^k = D_1^k$. Since $|D_1^k| > |D_2^k|$ and $|C| = |D_1^k|$, there is an $x \in C \setminus D_2^k$. This means there is an $x \in \{B_2^k \setminus B_1^k\} \setminus D_2^k$ with $\{B_1^k \setminus D_1^k\} \cup x \in \mathfrak{E}^k$. Now extend the latter set to a basis of $\mathfrak{E}^k$ by means of elements of B_1^k. It now follows that there is a $y \in D_1^k$ with $\{B_1^k \setminus y\} \cup x$ a basis of $\mathfrak{E}^k$, and thus that $\{B_1 \setminus y\} \cup x$ is a transversal of $\mathfrak{E}(I)$ whose complement contains $A_1 \cup y$ with $y \in A_2 \setminus A_1$. Thus $A_1 \cup y \in \underline{M}^*(\mathfrak{E}(I))$ which is what we set out to prove.

Suppose now that $\mathfrak{E}^i$ is a finite matroid $(i \in I)$. Let $F \subseteq E$ and suppose that for all $A \subset\subset F$, $E \setminus A$ contains a transversal of $\mathfrak{E}(I)$. We want to show that $E \setminus F$ contains a transversal of $\mathfrak{E}(I)$. This will mean that $\underline{M}^*(\mathfrak{E}(I))$ has finite character. For $i \in I$, $\mathfrak{E}^i_{E \setminus F}$ must have the same rank as $\mathfrak{E}^i$; otherwise if $A = \{x \in F : \{x\} \in \mathfrak{E}^i\}$, then A is a finite subset of F with $\mathfrak{E}^i_{E \setminus A}$ having rank smaller than $\mathfrak{E}^i$ and thus $E \setminus A$ cannot contain a transversal of $\mathfrak{E}(I)$. We can say therefore that $E \setminus F$ contains a transversal of $\mathfrak{E}(I)$ if and only if the family $(\mathfrak{E}^i_{E \setminus F} : i \in I)$ of finite matroids has a transversal. By Theorem 2.2 this is true if and only if $(\mathfrak{E}^i_{E \setminus F} : i \in K)$ has a transversal for all $K \subset\subset I$. Suppose there were a $K \subset\subset I$ such that $(\mathfrak{E}^i_{E \setminus F} : i \in K)$ does not have a transversal. Then $A = \{x \in F : \{x\} \in \mathfrak{E}^i\}$ for some $i \in K$ is a finite set and $E \setminus A$ does not contain a transversal of $(\mathfrak{E}^i_{E \setminus F} : i \in K)$. This is a contradiction. Thus $E \setminus F$ contains a transversal of $\mathfrak{E}(I)$.

A consequence of the last conclusion of the theorem is that every transversal of $\mathfrak{E}(I)$ contains a minimal transversal. Since $\underline{M}(\mathfrak{E}(I))$ is a finite character matroid when $\{i \in I : \{x\} \in \mathfrak{E}^i\}$ is a finite set $(x \in E)$, we can draw the following conclusion. Let $\mathfrak{E}(I) = (\mathfrak{E}^i : i \in I)$ be a family of finite matroids such that $\{i \in I : \{x\} \in \mathfrak{E}^i\}$ is a finite set for each $x \in E$ and suppose that $\mathfrak{E}(I)$ has a transversal. Then both $\underline{M}(\mathfrak{E}(I))$ and $\underline{M}^*(\mathfrak{E}(I))$ are finite character matroids, and given any transveral T of $\mathfrak{E}(I)$ there exists a minimal transversal T_1 and a maximal transversal T_2 with $T_1 \subseteq T \subseteq T_2$. The result for families of sets is that if a family of finite subsets of a set E with each element of E a member of only finitely many sets of the family has a transversal, then each transversal contains a minimal transversal and is contained in a maximal transversal.

Theorem 2.5 actually applies to a more general situation as the following theorem shows.

Theorem 2.8 <u>Let</u> $\mathcal{E}(I) = (\mathcal{E}^i : i \in I)$ <u>be a family of finite character matroids on a set</u> E. <u>Let</u> B <u>be a basis of the matroid</u> $\underline{M}(\mathcal{E}(I))$, <u>and let</u> $B = \sum_{i \in I} A_i$ <u>where</u> $A_i \in \mathcal{E}^i$ $(i \in I)$. <u>Let the family</u> $\tilde{\mathcal{E}}(I) = (\tilde{\mathcal{E}}^i : i \in I)$ <u>of finite character matroids be defined by</u> $\tilde{\mathcal{E}}^i = \mathcal{E}^i_{sp(A_i)}$ <u>where</u> $sp(A_i)$ <u>denotes the span of</u> A_i <u>in the matroid</u> $\mathcal{E}^i$ <u>on</u> E $(i \in I)$. <u>Then</u>

$$\underline{M}(\mathcal{E}(I)) = \underline{M}(\tilde{\mathcal{E}}(I)).$$

In proving the theorem suppose first that $\{i \in I : \{x\} \in \mathcal{E}^i\}$ is a finite set $(x \in E)$, so that both $\underline{M}(\mathcal{E}(I))$ and $\underline{M}(\tilde{\mathcal{E}}(I))$ are finite character matroids. Thus it is enough to show that if F is a finite set, $F \in \underline{M}(\mathcal{E}(I))$, then $F \in \underline{M}(\tilde{\mathcal{E}}(I))$. Suppose this is not so and choose F minimal with respect to the properties that $F \in \underline{M}(\mathcal{E}(I))$ but $F \notin \underline{M}(\tilde{\mathcal{E}}(I))$. Thus F is a circuit of $\underline{M}(\tilde{\mathcal{E}}(I))$. Since $F \in \underline{M}(\mathcal{E}(I))$ there must be an $x \in F \cap B$ such that $x \notin sp(A_k)$ for some $k \in K$. For each $y \in F \setminus B$, there is a circuit C_y of $\underline{M}(\mathcal{E}(I))$ with $y \in C_y \subseteq B \cup y$; each of these circuits must contain a circuit of $\underline{M}(\tilde{\mathcal{E}}(I))$. Since F is a circuit of $\underline{M}(\tilde{\mathcal{E}}(I))$, one of these latter circuits must contain x; otherwise repeated application of the circuit property produces a circuit of $\underline{M}(\tilde{\mathcal{E}}(I))$ contained entirely in B. Hence $B' = \{B \setminus x\} \cup y \in \underline{M}(\tilde{\mathcal{E}}(I))$. But then $B' \cup x = B \cup y$ is contained in $\underline{M}(\mathcal{E}(I))$ which is a contradiction. (This argument is adopted from an argument of Mason [19].)

In the general case, let $F \in \underline{M}(\mathcal{E}(I))$ so that $F = \sum_{i \in I} F_i$ where $F_i \in \mathcal{E}^i$ $(i \in I)$. Consider the family $\bar{\mathcal{E}}(I) = (\bar{\mathcal{E}}^i : i \in I)$ where $\bar{\mathcal{E}}^i = \mathcal{E}^i_{A_i \cup F_i}$ $(i \in I)$. Then for each $e \in E$, $|\{i \in I : e \in \bar{\mathcal{E}}^i\}| \leq 2$. Thus by what we have established above, $F \in \underline{M}(\mathcal{E}^i_{B_i} : i \in I)$ where B_i is the span of A_i in the matroid $\bar{\mathcal{E}}^i$ on E $(i \in I)$. Since B_i is contained in the span of A_i in the matroid $\mathcal{E}^i$ on E $(i \in I)$, it follows that $F \in \underline{M}(\mathfrak{E}(I))$, which is what was to be proved.

The theorem says, in particular, that if a family of finite character matroids has a maximal partial transversal, then it can be replaced by another family of finite character matroids which has a transversal with the collection of partial transversals of the individual families identical.

By means of Theorem 2.5, a theorem in generalized transversal theory which corresponds to a theorem of Hoffman and Kuhn [16] (proved only for finite families of subsets of a finite set), giving conditions for the existence of a transversal which intersects a given partition of the ground set within prescribed bounds, can be obtained. A more comprehensive result, including a transfinite extension of this theorem, was proved in [4]. The corresponding theorem in generalized transversal theory was proved in [8] and this, with a slight change, follows.

Theorem 2.9 <u>Let</u> $\mathfrak{E}(I) = (\mathfrak{E}^i : i \in I)$ <u>be a family of finite matroids on</u> E <u>where the sets</u> $\{i \in I : \{x\} \in \mathfrak{E}^i\}$ <u>are assumed to be finite</u> $(x \in E)$. <u>Let</u> $E = \sum_{j \in J} E_j$ <u>be a partition of</u> E <u>into finite sets, and let integers</u> c_j, d_j <u>with</u> $0 \leq c_j \leq d_j \leq |E_j|$ $(j \in J)$ <u>be given. Then</u> $\mathfrak{E}(I)$ <u>has a transversal</u> T <u>with</u>

$$c_j \leq |T \cap E_j| \leq d_j \qquad (j \in J)$$

<u>if and only if</u>

(i) $$\sum_{j \in J} \min\{d_j, |A \cap E_j|\} \geq \sum_{i \in I} r_A^i(A) \qquad (A \subset\subset E)$$

(ii) $$\sum_{i \in I} r^i(F) + |E(K) \setminus F| \geq \sum_{j \in K} c_j \qquad (K \subset\subset J, \ F \subseteq E(K)).$$

A special case of the above Theorem gives necessary and sufficient conditions in order that the ground set E be a transversal of $\mathfrak{E}(I)$. This results when E is partitioned into singletons and all the c's are chosen to be 1. (Another special case gives necessary and sufficient conditions that $\mathfrak{E}(I)$ have a transversal containing a prescribed set.)

Theorem 2.10 Let $\mathcal{E}(I) = (\mathcal{E}^i : i \in I)$ be a family of finite matroids on E with $\{i \in I : \{x\} \in \mathcal{E}^i\}$ a finite set $(x \in E)$. Then E is a transversal of $\mathcal{E}(I)$ if and only if

$$\sum_{i \in I} r^i(A) \geq |A| \geq \sum_{i \in I} r^i_A(A) \qquad (A \subset\subset E).$$

Indeed the first inequality for all $A \subset\subset E$ is equivalent to $\mathcal{E}(I)$ having a transversal while the second inequality for all $A \subset\subset E$ is equivalent to E being a partial transversal of $\mathcal{E}(I)$. Both together are then equivalent to E being a transversal of $\mathcal{E}(I)$.

Rado ([27], [28]) has generalized Theorem 2.1 in an extremely powerful way. This theorem which is of fundamental importance in transversal theory is as follows.

Theorem 2.10 (Rado) Let $\mathfrak{A}(I) = (A_i : i \in I)$ be a family of subsets of a set E, and let $\mathcal{E}$ be a finite character matroid on E with rank function r. If I is an infinite set, assume A_i is a finite set $(i \in I)$. Then $\mathfrak{A}(I)$ has a transversal T with $T \in \mathcal{E}$ if and only if

$$r(A(K)) \geq |K| \qquad (K \subset\subset I).$$

The corresponding theorem in generalized transversal theory is proved in [2] and [8]. (See also [6].)

Theorem 2.11 Let $\mathcal{E}(I) = (\mathcal{E}^i : i \in I)$ be a family of rank-finite matroids on E, and let $\mathcal{E}$ be a finite character matroid on E with rank function r.

(i) If $|I| < \infty$, there exists a transversal T of $\mathcal{E}(I)$ with $T \in \mathcal{E}$ if and only if

$$r(A) \geq \sum_{i \in I} r^i_A(A) \qquad (A \subseteq E).$$

(ii) <u>If</u> I <u>is arbitrary and</u> $\mathcal{E}^i$ <u>is a finite matroid</u> $(i \in I)$, <u>there exists a transversal</u> T <u>of</u> $\mathcal{E}(I)$ <u>with</u> $T \in \mathcal{E}$ <u>if and only if</u>

$$r(A) \geq \sum_{i \in I} r^i_A(A) \qquad (A \subset\subset E).$$

There is one more important theorem in transversal theory and its corresponding theorem in generalized transversal theory that I wish to discuss. The theorem in transversal theory is the Ford-Fulkerson theorem [18] which gives necessary and sufficient conditions for two finite families of sets to have a common transversal. A transfinite analog, along with more general results, is proved by me in [7]. If $\mathcal{E}(I) = (\mathcal{E}^i : i \in I)$ and $\mathcal{F}(J) = (\mathcal{F}^j ; j \in J)$ are two families of matroids on a set E, then a set $T \subseteq E$ is a <u>common transversal</u> of $\mathcal{E}(I)$ and $\mathcal{F}(J)$ provided it is a transversal of both. To obtain conditions for two families of matroids to have a common transversal, one needs the following result [8].

Theorem 2.12 <u>Let</u> $\mathcal{E}(I) = (\mathcal{E}^i : i \in I)$ <u>and</u> $\mathcal{F}(J) = (\mathcal{F}^j : j \in J)$ <u>be two families of rank-finite matroids on a set</u> E. <u>Then</u> $\mathcal{E}(I)$ <u>and</u> $\mathcal{F}(J)$ <u>have a common transversal if and only if there is a transversal of</u> $\mathcal{E}(I)$ <u>which is a partial transversal of</u> $\mathcal{F}(J)$ <u>and there exists a transversal of</u> $\mathcal{F}(J)$ <u>which is a partial transversal of</u> $\mathcal{E}(I)$.

From Theorems 2.5, 2.10 and 2.12 we can obtain conditions for two families of matroids to have a common transversal [8].

Theorem 2.13. <u>Let</u> $\mathcal{E}(I) = (\mathcal{E}^i : i \in I)$ <u>and</u> $\mathcal{F}(J) = (\mathcal{F}^j : j \in J)$ <u>be two families of rank-finite matroids on</u> E <u>with rank functions</u> r^i $(i \in I)$ <u>and</u> ρ^j $(j \in J)$, <u>respectively</u>.

(i) <u>If</u> $\sum_{i \in I} r^i(E) = \sum_{j \in J} \rho^j(E) < \infty$; <u>then</u> $\mathcal{E}(I)$ <u>and</u> $\mathcal{F}(J)$ <u>have a common transversal if and only if</u>

$$|F| \geq \sum_{i \in I} r^i_A(A) - \sum_{j \in J} \rho^j(A \setminus F) \qquad (F \subset\subset A \subseteq E).$$

(ii) If I and J are arbitrary, $\mathcal{E}^i$ $(i \in I)$ and $\mathcal{F}^j$ $(j \in J)$ are finite matroids, $\{i \in I : \{e\} \in \mathcal{E}^i\}$ and $\{j \in J : \{e\} \in \mathcal{F}^j\}$ are finite sets $(e \in E)$, then $\mathcal{E}(I)$ and $\mathcal{F}(J)$ have a common transversal if and only if

$$|F| \geq \max\{\sum_{i \in I} r^i_A(A) - \sum_{j \in J} \rho^j(A \setminus F), \sum_{j \in J} \rho^i_A(A) - \sum_{i \in I} r^i(A \setminus F)\}$$

$$(F \subseteq A \subset\subset E).$$

3. INCIDENCE RELATIONS AND FAMILIES OF MATROIDS

Thus far I have been concerned with one or two families of matroids on the same set E, and I have indicated a number of results that can be proved. Most of these results can be extended to the setting where there is a family of matroids on each of two sets, X and Y, which are connected by an incidence relation. The key to obtaining theorems in this setting, especially when X and Y are not assumed to be finite sets and the families of matroids are not finite families, is more often than not a mapping or linking theorem. If $\Delta \subseteq X \times Y$ is an incidence relation between X and Y, then sets $A \subseteq X$ and $B \subseteq Y$ are linked (or A is linked to B, or B is linked to A) if there is a bijection $\theta : A \to B$ with $(a, \theta(a)) \in \Delta$ $(a \in A)$. Thus each element of A is incident with its image. The following linking theorem which contains as special cases many other linking theorems (see [1], [3], [25]) is proved in [8].

Theorem 3.1 Let X and Y be sets and $\Delta \subseteq X \times Y$ an incidence relation between X and Y. Let $\mathcal{E}(I) = (\mathcal{E}^i : i \in I)$ be a family of rank-finite matroids on X and let $\mathcal{F}(J) = (\mathcal{F}^j : j \in J)$ be a family of rank-finite matroids on Y. Let $\bar{I} \subseteq I$ and $\bar{J} \subseteq J$. The following two statements ((i) and (ii)) are equivalent. †

(i) There is a partial transversal $\tilde{X}$ of $\mathcal{E}(I)$ with $\tilde{X} = \sum_{i \in I} \tilde{X}_i$ with $\tilde{X}_i$ a

† In [8] the matroids $\mathcal{E}^i$ $(i \in I)$ and ^{j} $(j \in J)$ were assumed to be finite matroids, but the same proof works for rank-finite matroids.

basis of $\mathfrak{E}^i$ $(i \in \bar{I})$ and $\tilde{X}_i \in \mathfrak{E}^i$ $(i \in I \setminus \bar{I})$, and there is a partial transversal $\tilde{Y}$ of $\mathfrak{F}(J)$ with $\tilde{Y} = \sum_{j \in J} \tilde{Y}_j$ with $\tilde{Y}_j$ a basis of $\mathfrak{F}^j$ $(j \in \bar{J})$ and $\tilde{Y}_j \in \mathfrak{F}^j$ $(j \in J \setminus \bar{J})$ such that $\tilde{X}$ and $\tilde{Y}$ are linked.

(ii) (a) There exists a transversal of $\mathfrak{E}(\bar{I})$ which is linked to a partial transversal of $\mathfrak{F}(J)$.

(b) There exists a transversal of $\mathfrak{F}(\bar{J})$ which is linked to a partial transversal of $\mathfrak{E}(I)$.

If $\bar{I} = I$, $\bar{J} = J$, the following corollary results.

Corollary 3.2 There exists a transversal of $\mathfrak{E}(I)$ and a transversal of $\mathfrak{F}(J)$ which are linked if and only if there is a transversal of $\mathfrak{E}(I)$ which is linked to a partial transversal of $\mathfrak{F}(J)$ and there is a transversal of $\mathfrak{F}(J)$ which is linked to a partial transversal of $\mathfrak{E}(I)$.

Theorem 3.1 can be used to obtain a more comprehensive result; for details, see [8].

If X and Y are sets which are connected by an incidence relation Δ, it may be assumed without loss in generality that $X \cap Y = \emptyset$. This is because we may replace Y by a "copy" Y' with $X \cap Y' = \emptyset$ and carry over any structures on Y to Y'. If this convention is agreed to, the following notation is not ambiguous. For $x \in X$, we let $\Delta(x) = \{y \in Y : (x, y) \in \Delta\}$, and for $A \subseteq X$ we let $\Delta(A) = \bigcup_{x \in A} \Delta(x)$. For $y \in Y$ and $B \subseteq Y$, $\Delta(y)$ and $\Delta(B)$ are defined in a similar way.

The following theorem generalizes Theorem 2.11. Its results are contained in [2] and [8].

Theorem 3.3 Let X and Y be sets with $\Delta \subseteq X \times Y$ an incidence relation between X and Y. Let $\mathfrak{E}(I) = (\mathfrak{E}^i : i \in I)$ be a family of rank-finite matroids on X, and let $\mathfrak{E}$ be a matroid on Y with rank function ρ.

(i) If $|I| < \infty$, then there is a transversal T of $\mathfrak{E}(I)$ which is linked to a set $F \in \mathfrak{E}$ if and only if

$$|A| \geq \sum_{i \in I} r^i_A(A) \qquad (A \subset\subset X)$$

$$\rho(\Delta(A)) \geq \sum_{i \in I} r^i_A(A) . \qquad (A \subseteq X) .$$

(ii) If I is arbitrary, $\mathfrak{E}^i$ is a finite matroid $(i \in I)$, and $\Delta(x)$ is a finite set $(x \in X)$, then there is a transversal T of $\mathfrak{E}(I)$ which is linked to a set $F \in \mathfrak{E}$ if and only if

$$\min \{|A|, \rho(\Delta(A))\} \geq \sum_{i \in I} r^i_A(A) \qquad (A \subset\subset X)$$

The next result generalizes Theorem 2.13 and is a consequence of results proved in [8].

Theorem 3.4 Let X and Y be sets and $\Delta \subseteq X \times Y$ an incidence relation between X and Y with $\Delta(x)$ and $\Delta(y)$ finite sets $(x \in X,\ y \in Y)$. Let $\mathfrak{E}(I) = (\mathfrak{E}^i : i \in I)$ be a family of finite matroids on X and $\mathfrak{F}(J) = (\mathfrak{F}^j : j \in J)$ a family of finite matroids on Y with rank functions denoted, respectively, by r^i $(i \in I)$ and ρ^j $(j \in J)$. Assume that $\{i \in I : \{x\} \in \mathfrak{E}^i\}$ and $\{j \in J : \{y\} \in \mathfrak{F}^j\}$ are finite sets $(x \in X,\ y \in Y)$. Then there is a transversal of $\mathfrak{E}(I)$ which is linked to a transversal of $\mathfrak{F}(J)$ if and only if

$$|A| \geq \sum_{i \in I} r^i_A(A) \qquad (A \subset\subset X)$$

$$|B| \geq \sum_{j \in J} \rho^j_B(B) \qquad (B \subset\subset X)$$

$$|\Delta(A) \setminus P| + \sum_{j \in J} \rho^j(P) \geq \sum_{i \in I} r^i_A(A) \qquad (A \subset\subset X,\ P \subseteq \Delta(A))$$

$$|\Delta(B)\setminus Q| + \sum_{i\in I} r^i(Q) \geq \sum_{j\in J} \rho^j_B(B) \qquad (B \subset\subset Y,\ Q \subseteq \Delta(B)).$$

This theorem follows from a double application of Theorem 3.3 using the finite character matroids $\underline{M}(\mathcal{E}(I))$ and $\underline{M}(\mathcal{F}(J))$, respectively, and the rank formulas in Theorem 2.5. Corollary 3.2 then completes the proof.

If all the finitary restrictions are removed in Theorem 3.4, the following result can still be obtained.

Theorem 3.5 Let X and Y be sets and $\Delta \subseteq X \times Y$ an incidence relation between X and Y. Let $\mathcal{E}(I) = (\mathcal{E}^i : i \in I)$ and $\mathcal{F}(J) = (\mathcal{F}^j : j \in J)$ be families of matroids on X and Y, respectively. Let t be a nonnegative integer. Then there exist partial transversals of $\mathcal{E}(I)$ and $\mathcal{F}(J)$ with cardinality t which are linked if and only if

$$|A\setminus P| + \sum_{i\in I} r^i(P) + |\Delta(A)\setminus Q| + \sum_{j\in J} \rho^j(Q) \geq t \qquad (P \subseteq A \subseteq X,\ Q \subseteq \Delta(A)).$$

In the special case when $|I| = 1$ with $\mathcal{E}(I)$ consisting of one matroid $\mathcal{E}$ with rank function r and when $|J| = 1$ with $\mathcal{F}(J)$ consisting of one matroid $\mathcal{F}$ with rank function ρ, the above condition reduces to

$$r(A) + \rho(\Delta(A)) \geq t \qquad (A \subseteq X).$$

This result is proved in [2] (in the finite situation) and in [7] (in the general situation); see also [31] for the finite situation. Theorem 3.5 follows from this special case using Theorem 2.5.

4. TRANSVERSALS OF FAMILIES OF MATROIDS

Given a family $\mathcal{E}(I) = (\mathcal{E}^i : i \in I)$ of matroids on a set E with I an infinite index set, in writing down conditions for $\mathcal{E}(I)$ to have a transversal it was assumed that $\mathcal{E}^i$ was a finite matroid $(i \in I)$. This corresponds to the situation for infinite families of sets when in writing down conditions for a transversal it was assumed that the sets were finite sets. Rado and Jung [29] have obtained conditions for an infinite family of sets to have a transversal if only one set of the family could be infinite. This result was generalized by Scrimger and me [9] to allow for the possibility of finitely many sets of the family being infinite. Results of this type can also be obtained for families of matroids, and I shall indicate how. First I require a result which gives as a special case necessary and sufficient conditions for a family of finite matroids to have a transversal with "defect" at most d. This result can then be used to evaluate the rank function of the finite character matroid $M^*(\mathfrak{E}(I))$ where $\mathfrak{E}(I)$ is a family of finite matroids. In some of the results that follow, if I is a finite set, it could be assumed that the matroids involved are only rank-finite rather than finite; but I shall forego this generality.

Theorem 4.1 <u>Let</u> $\mathfrak{E}(I) = (\mathfrak{E}^i : i \in I)$ <u>be a family of finite matroids on</u> E. <u>Let</u> $(t_i : i \in I)$ <u>be a family of nonnegative integers with</u> $t_i \geq r^i(E)$ $(i \in I)$, <u>and let</u> d <u>be a nonnegative integer</u>. <u>Then</u> $\mathfrak{E}(I)$ <u>has a partial transversal</u> A <u>where</u> $A = \sum_{i \in I} A_i$ <u>with</u> $A_i \in \mathfrak{E}^i$ $(i \in I)$ <u>and</u> $d \geq \sum_{i \in I} (t_i - |A_i|)$ <u>if and only if</u>

$$|F| + d \geq \sum_{i \in I} (t_i - r^i(E \setminus F)) \qquad (F \subset\subset E).$$

Observe that if the above condition is satisfied with $F = \emptyset$, then $d \geq \sum_{i \in I} (t_i - r^i(E))$. In particular, $d \geq t_i - r^i(E)$ $(i \in I)$ with $t_i = r^i(E)$ for all but at most finitely many $i \in I$.

To prove the conditions given in the theorem are sufficient for the existence of the partial transversal of the desired type (they are surely necessary), it is only necessary to consider the family $(\mathfrak{F}^i : i \in I)$ of matroids where $\mathfrak{F}^i = (\mathfrak{E}^i \oplus \mathcal{P}(D))_{t_i}$ $(i \in I)$ where D is a set with $D \cap E = \emptyset$, $|D| = d$ and $\mathcal{P}(D)$ denotes the collection (matroid) of all subsets of D. Thus the finite matroid $\mathfrak{F}^i$ is obtained by taking the direct sum of $\mathfrak{E}^i$ and $\mathcal{P}(D)$ and then truncating at t_i $(i \in I)$. The desired partial transversal of $\mathfrak{E}(I)$ exists if and only if this newly constructed family of finite matroids has a transversal. Theorem 2.2 can now be applied to obtain the desired result.

If $t_i = r^i(E)$ $(i \in I)$, the theorem states that $\mathfrak{E}(I)$ has a partial transversal with "defect" at most d if and only if

$$|F| + d \geq \sum_{i \in I} r^i_F(F) \qquad (F \subset\subset E).$$

This extends a theorem in transversal theory to generalized transversal theory (see [29] for this result in transversal theory).

From Theorem 4.1 a formula for the rank function of the matroid $\underline{M}^*(\mathfrak{E}(I))$ of Theorem 2.7 can be obtained.

Theorem 4.2 <u>Let</u> $\mathfrak{E}(I) = (\mathfrak{E}^i : i \in I)$ <u>be a family of finite matroids on</u> E <u>which has a transversal</u>. <u>Let</u> r^* <u>denote the rank function of the finite character matroid</u> $M^*(\mathfrak{E}(I))$.

(i) <u>If</u> $A \subset\subset E$, <u>then</u>

$$r^*(A) = \min_{F \subset\subset E \setminus A} \{|A| + |F| - \sum_{i \in I} r^i_{A \cup F}(A \cup F)\}$$

(ii) <u>If</u> $A \subseteq E$, <u>then</u>

$$r^*(A) = \sup_{X \subset\subset A} \ \min_{F \subset\subset E \setminus X} \{|X| + |F| - \sum_{i \in I} r^i_{X \cup F}(X \cup F)\}.$$

To prove (i), let $A \subset\subset E$. If d is a nonnegative integer, then $r^*(A) \geq |A| - d$ provided the family $(\mathfrak{E}^i_{E \setminus A} : i \in I)$ of finite matroids has a partial transversal T with $T = \sum_{i \in I} T_i$ where $T_i \in \mathfrak{E}^i$ $(i \in I)$ and

$$d \geq \sum_{i \in I} (r^i(E) - |T_i|).$$

(It is not difficult to show that, since $\mathfrak{E}(I)$ has a transversal, a partial transversal with "defect" at most d can be enlarged to a transversal by the addition of at most d elements). By Theorem 5.1 this is so if and only if

$$|F| + d \geq \sum_{i \in I} r^i(E) - r^i(E \setminus A \setminus F) \qquad (F \subset\subset E \setminus A),$$

and from this (i) easily follows. Part (ii) follows from (i) since if $A \subseteq E$, $r^*(A) = \sup_{X \subset\subset A} r^*(X)$. (Recall that the rank function of a matroid takes on values $0, 1, \ldots, \infty$; $r^*(A) = \infty$ if $r^*(X)$ is unbounded as X ranges over the finite subsets of A).

If $\mathcal{E}(I) = (\mathcal{E}^i : i \in I)$ and $\mathcal{F}(J) = (\mathcal{F}^j : j \in J)$ are two families of matroids, then $\mathcal{E}(I) \dot{+} \mathcal{F}(J) = (\mathcal{G}^k : k \in I \cup J)$ where $\mathcal{G}^k = \mathcal{E}^k$ if $k \in I$ and $\mathcal{G}^k = \mathcal{F}^k$ if $k \in J$. Here we are assuming, without loss in generality, that $I \cap J = \emptyset$.

Theorem 4.3 <u>Let</u> $\mathcal{E}(I) = (\mathcal{E}^i : i \in I)$ <u>be a family of finite matroids on</u> E <u>with rank functions</u> r^i $(i \in I)$, <u>and let</u> $\mathcal{F}(J) = (\mathcal{F}^j : j \in J)$ <u>be a finite family of rank-finite matroids on</u> E <u>with rank functions</u> ρ^j $(j \in J)$. <u>Then</u> $\mathcal{E}(I) \dot{+} \mathcal{F}(J)$ <u>has a transversal if and only if</u>

$$\text{(i)} \quad |A| \geq \sum_{i \in I} r^i_A(A) \qquad (A \subset\subset E)$$

$$\text{(ii)} \quad \sup_{X \subset\subset A} \; \min_{F \subset\subset E \setminus X} \Big\{ |X| + |F| - \sum_{i \in I} r^i_{X \cup F}(X \cup F) \Big\} \geq \sum_{j \in J} \rho^j_A(A) \qquad (A \subseteq E).$$

From Theorem 2.3, (i) is equivalent to $\mathcal{E}(I)$ having a transversal. It then follows that $\mathcal{E}(I) + \mathcal{F}(J)$ has a transversal if and only if $\mathcal{F}(J)$ has a transversal T with $T \in \underline{M}^*(\mathcal{E}(I))$. By Theorem 2.11(i) and the evaluation of the rank function of $\underline{M}^*(\mathcal{E}(I))$ as given in Theorem 5.2(ii), this is equivalent to (ii) above.

Let me now take a different approach. Suppose $\mathcal{E}(I) = (\mathcal{E}^i : i \in I)$ is a family of finite character matroids on E for which one is interested in determining whether or not it possesses a transversal. Can the structure of the matroid $\underline{M}(\mathcal{E}(I))$ be useful in this determination? This approach was used in [10] where it was determined that if $\mathfrak{A}(I)$ is a family of sets for which the matroid $\underline{M}(\mathfrak{A}(I))$ has a basis and no coloops, then $\mathfrak{A}(I)$ has a transversal. Analogous results can be obtained in generalized transversal theory, and I shall indicate some of these in the remaining space. The first thing needed is a lemma.

Lemma 4.4 Let $\mathfrak{E}(I) = (\mathfrak{E}^i : i \in I)$ be a family of finite character matroids on E such that $\underline{M}(\mathfrak{E}(I))$ is coloop-free and has at least one basis. If B is a basis of $\underline{M}(\mathfrak{E}(I))$, then given $x \in B$ there exists $y \in E \setminus B$ such that $\{B \setminus x\} \cup y \in \underline{M}(\mathfrak{E}(I))$.

($\{B \setminus x\} \cup y$ must be a basis of $\underline{M}(\mathfrak{E}(I))$ but this will not be needed.) The proof will be by contradiction. Let B be a basis of $\underline{M}(\mathfrak{E}(I))$ and suppose there is $x \in B$ such that $\{B \setminus x\} \cup y \notin \underline{M}(\mathfrak{E}(I))$ for all $y \in E \setminus B$. Let $B = \sum_{i \in I} B_i$ where $B_i \in \mathfrak{E}^i$ $(i \in I)$. Let $A \in \underline{M}(\mathfrak{E}(I))$, $A = \sum_{i \in I} A_i$ where $A_i \in \mathfrak{E}^i$ $(i \in I)$, and suppose that $A \cup x \notin \underline{M}(\mathfrak{E}(I))$. Consider the family $\tilde{\mathfrak{E}}(I) = (\mathfrak{E}^i_{A_i \cup B_i} : i \in I)$ of finite character matroids. As in similar arguments, the matroid $\underline{M}(\tilde{\mathfrak{E}}(I))$ has finite character. Moreover $A, B \in \underline{M}(\tilde{\mathfrak{E}}(I))$ with B a basis and $A \cup x \notin \underline{M}(\tilde{\mathfrak{E}}(I))$. Thus for all $y \in A \setminus B$ the unique circuit of $\underline{M}(\tilde{\mathfrak{E}}(I))$ contained in $B \cup y$ does not contain x (if it did $\{B \setminus x\} \cup y \in \underline{M}(\tilde{\mathfrak{E}}(I)) \subseteq \underline{M}(\mathfrak{E}(I))$. By repeated application of the circuit property, a circuit C is produced with $x \in C \subseteq B$; but this contradicts the fact that $B \in \underline{M}(\tilde{\mathfrak{E}}(I))$. Hence $A \cup x \in \underline{M}(\mathfrak{E}(I))$. Since this is true for all $A \in \underline{M}(\mathfrak{E}(I))$, x is a coloop which contradicts an assumption of the lemma.

Combining this lemma with Theorem 2.8 leads to the following result.

Theorem 4.5 Let $\mathfrak{E}(I) = (\mathfrak{E}^i : i \in I)$ be a family of finite character matroids on E such that the matroid $\underline{M}(\mathfrak{E}(I))$ is a coloop-free matroid with at least one basis. Then $\mathfrak{E}(I)$ has a transversal.

I argue as follows. Let B be a basis of $\underline{M}(\mathfrak{E}(I))$ so that, in particular,

$B = \sum_{i \in I} B_i$ where $B_i \in \mathcal{E}^i$ $(i \in I)$. Suppose for some $k \in I$ that B_k were not a basis of $\mathcal{E}^k$. Then there exists $x \in B \setminus B_k$ such that $B_k \cup x \in \mathcal{E}^k$. By Lemma 4.4 there exists $y \in E \setminus B$ with $A = \{B \setminus x\} \cup y \in \underline{M}(\mathcal{E}(I))$. By Theorem 2.8, $A \in \underline{M}(\tilde{\mathcal{E}}(I))$ where $\tilde{\mathcal{E}}(I) = (\mathcal{E}^i_{sp(B_i)} : i \in I)$. But then $A \cup x \in \underline{M}(\tilde{\mathcal{E}}(I))$; since $A \cup x = B \cup y$ where $y \notin B$, this contradicts the maximality condition on B.

The above proof establishes more than that stated in the theorem; namely, if $B = \sum_{i \in I} B_i$ where $B_i \in \mathcal{E}^i$, then B_i must be a basis of $\mathcal{E}^i$ $(i \in I)$. From this the following theorem can be obtained.

Theorem 4.6 Let $\mathcal{E}(I) = (\mathcal{E}^i : i \in I)$ be a family of finite character matroids on E such that the matroid $\underline{M}(\mathcal{E}(I))$ has a basis and only a finite number of coloops. Then $\mathcal{E}(I)$ has a transversal if and only if

$$|A| \geq \sum_{i \in I} r^i_A(A) \qquad (A \subset\subset E).$$

The condition is surely necessary for $\mathcal{E}(I)$ to have a transversal. Suppose now it is satisfied. Let F be the set of coloops of $\underline{M}(\mathcal{E}(I))$ so that $|F| < \infty$. Let B be a basis of $\underline{M}(\mathcal{E}(I))$ and consider $\tilde{\mathcal{E}}(I) = (\mathcal{E}^i_{E \setminus F} : i \in I)$. Then $\underline{M}(\tilde{\mathcal{E}}(I))$ is a coloop-free matroid and has $B \setminus F$ as basis. Thus by Theorem 4.5 (and the remark following its proof), $B \setminus F = \sum_{i \in I} A_i$ where A_i is a basis of $\mathcal{E}^i_{E \setminus F}$ and if $B \setminus F = \sum_{i \in I} A'_i$ with $A'_i \in \mathcal{E}^i_{E \setminus F}$ $(i \in I)$, then A'_i is a basis of $\mathcal{E}^i_{E \setminus F}$ $(i \in I)$. Thus $\mathcal{E}(I)$ has a transversal if and only if the family $(\mathcal{E}^i_{\otimes F} : i \in I)$ of matroids on the finite set F has a transversal. By taking $A = F$ in the condition in the theorem, there results $|F| \geq \sum_{i \in I} r^i_F(F)$; thus $r^i_F(F) = 0$ except for finitely many $i \in I$, so that the family $(\mathcal{E}^i_{\otimes F} : i \in I)$ is effectively a finite family. By Theorem 2.2(i), it has a transversal if and only if

$$|A| \geq \sum_{i \in I} (r^i_F)_A(A) \qquad (A \subseteq F).$$

Since $(r_F^i)_A(A) = r_A^i(A)$ $(i \in I)$, this proves the theorem.

The problem of finding necessary and sufficient conditions in order that a family $(\mathcal{E}^i : 1 \leq i \leq n)$ of finite character matroids have a transversal is, to my knowledge, unsolved, except under the circumstances given in Theorem 4.6.

References

[1] Banach, S., Un théoremè sur les transformations biunivoques. Fundamenta Math. 6 (1927), 236-239.

[2] Brualdi, R. A., Symmetrization of R. Rado's theorem on independent representatives. Unpublished paper (1967).

[3] __________, An extension of Banach's mapping theorem. Proc. Amer. Math. Soc. 20 (1969), 520-526.

[4] __________, A very general theorem on systems of distinct represenatives. Trans. Amer. Math. Soc. 140 (1969), 149-160.

[5] __________, Comments on bases in dependence structures. Bull. Australian Math. Soc. 1 (1969), 161-167.

[6] __________, Admissible maps between dependence spaces. Proc. London Math. Soc. (1970).

[7] __________, A general theorem concerning common transversals. To be published in the Proceedings of the Oxford Conference on Combinatorial Mathematics and its Applications, 1969 (Academic Press).

[8] __________, On families of finite independence structures. To be published in Proc. London Math. Soc.

[9] Brualdi, R. A. and Scrimger, E. B., Exchange systems, matchings and transversals. J. Combinatorial Theory 5 (1968), 244-257.

[10] Brualdi, R. A. and Mason, J. H., Transversal matroids and Hall's theorem. Submitted for publication.

[11] Dlab, V., The role of the "finite character property" in the theory of dependence. Comm. Math. Univ. Cardinal 6 (1965), 97-104.

[12] Edmonds, J., Matroid partition. Lectures in Applied Mathematics Vol. 1 (Mathematics of the Decision Sciences).

[13] Edmonds, J. and Fulkerson, D. R., Transversal and matroid partition. J. Res. Nat'l. Bur. Standards. 69B (1965), 147-153.

[14] Hall, P., On representatives of subsets. J. London Math. Soc. 10 (1935), 26-30.

[15] Hall, M., Jr., Distinct representatives of subsets. Bull. Amer. Math. Soc. 54 (1948), 922-926.

[16] Hoffman, A. J. and Kuhn, H. W., On systems of distinct representatives. Annals of Math. Studies No. 38, Princeton Univ. Press (1956).

[17] ____________, Systems of distinct representatives and linear algebra. Amer. Math. Monthly 63 (1956), 455-460.

[18] Ford, L. R. and Fulkerson, D. R., Network flow and systems of representatives. Canad. J. Math. 10 (1958), 78-84.

[19] Mason, J. H., A characterization of transversal spaces. To be published.

[20] Mendelsohn, N. S. and Dulmage, A. L., Some generalizations of the problem of distinct representatives. Canad. J. Math. 10 (1958), 230-241.

[21] Mirsky, L., Transversals of subsets. Quart. J. Math. (Oxford) (2), 17 (1966), 58-60.

[22] Mirsky, L. and Perfect, H., Systems of representatives. J. Math. Anal. and Applics. 15 (1966), 520-568.

[23] ____________, Applications of the notion of independence to problems of combinatorial analysis. J. Combinatorial Theory 2 (1967), 327-357.

[24] Nash-Williams, C. St. J. A., An application of matroids to graph theory. Proc. Symp. Rome, Dunod (1968), 263-265.

[25] Perfect, H. and Pym, J. S., An extension of Banach's mapping theorem with applications to problems concerning common representatives. Proc. Cambridge Philos. Soc. 62 (1966), 187-192.

[26] Pym, J. S. and Perfect, H., Submodular functions and independence structures. J. Math. Anal. and Applics. 30 (1970), 1-31.

[27] Rado, R., A theorem on independence relations. Quart. J. Math. (Oxford) 13 (1942), 83-89.

[28] ____________, Axiomatic treatment of rank in infinite sets. Canad. J. Math. 1 (1949), 337-343.

[29] ____________, Note on the transfinite case of Hall's theorem on representatives. J. London Math. Soc. 42 (1967), 321-324.

[30] Tutte, W. T., Lectures on matroids. J. Res. Nat'l. Bur. Standards 69B (1965), 1-47.

[31] Welsh, D. J. A., On matroid theorems of Edmonds and Rado. J. London Math. Soc. 2 (1970), 251-256.

[32] Whitney, H., On the abstract properties of linear dependence. Amer. J. Math. 57 (1935), 509-533.

AN EXTREMAL PROBLEM FOR NON-SEPARABLE MATROIDS

George W. Dinolt[1]

Abstract

Let E be a finite set, $\mathcal{E} \subseteq \mathcal{P}(E)$ a matroid on $\mathcal{E}$. $\mathcal{E}$ is non-separable if every pair of elements of E are in a common circuit (minimal member of $\mathcal{P}(E) \setminus \mathcal{E}$). The following two theorems are proved:

1. Every nonseparable matroid of rank r and cardinality n must have at least $r(n-r) + 1$ bases (maximal members of $\mathcal{E}$). There is for each r and n, a unique matroid with precisely this many bases.

2. If $\mathcal{E}$ is a nonseparable matroid of rank r on n elements containing more than $r(n-r) + 1$ bases, then $\mathcal{E}$ must contain at least $2(r-1)(n-r-1) + 1 + r$ bases if $2r \leq n$ and $(2r-1)(n-r-1) + 2$ bases if $2r \geq n$.

1. Introduction

We will always assume, in the following discussion, that E is some fixed finite set with $|E| = n < \infty$. Let $\mathcal{E} \subseteq \mathcal{P}(E)$ satisfy:

(E_1) $\emptyset \in \mathcal{E}$,

(E_2) If $A \in \mathcal{E}$ and $B \subseteq A$ then $B \in \mathcal{E}$.

(E_3) If $A, B \in \mathcal{E}$ are such that $|A| = k$ and $|B| = k + 1$, then there exists a $b \in B \setminus A$ such that $A \cup b \in \mathcal{E}$.

The elements of $\mathcal{E}$ are called <u>independent sets</u> and the pair $(E, \mathcal{E})$ or sometimes just $\mathcal{E}$ is called a <u>matroid</u> or <u>independence structure</u> on E.

The collection of minimal members of $\mathcal{P}(E) \setminus \mathcal{E}$ is the collection of <u>circuits</u> of $\mathcal{E}$. The elements of $\mathcal{C}$ satisfy:

---

[1] The research for this paper was conducted while the author was a Visiting Post Graduate Student at the University of Sheffield, Sheffield, England.

(C_1) No proper subset of a circuit is a circuit.

(C_2) If $C_1, C_2 \in \mathcal{C}$ are such that there is an $x \in C_1 \cap C_2$ and $y \in C_1 \setminus C_2$, then there is a $C_3 \in \mathcal{C}$ satisfying $y \in C_3 \subseteq (C_1 \cup C_2) \setminus x$.

The maximal members of $\mathcal{E}$ are called the bases of $(E, \mathcal{E})$. The symbol $\mathcal{B}$ will denote the collection of maximal members of $\mathcal{E}$. It is well known that for any B_1 and $B_2 \in \mathcal{B}$, $|B_1| = |B_2|$.

We can also define a rank function $r : \mathcal{P}(E) \to \mathbb{Z}^+ \cup \{0\}$ by $r(A) = \max\{|K| : K \subseteq A, K \in \mathcal{E}\}$ for any $A \subseteq E$. We will always assume that $r(E) = r$. It is interesting to note that if $A \subseteq \mathcal{E}$ then $r(A) = |A|$ and that $r(E) = r(B) = r$ for any $B \in \mathcal{B}$.

It can be shown that if one has a collection of subsets of E which satisfy properties (C_1) and (C_2), then it is possible to construct an independence structure $\mathcal{E}$ and hence the set $\mathcal{B}$ and the rank function r. One can also characterize the collection of bases $\mathcal{B}$ or the rank function r in a similar way. For a complete discussion of these matters, see Whitney's classic paper [5] and [1, 4]. In the sequel we assume that matroids are completely characterized by any of the following: independent sets, bases, circuits, or rank functions.

We can define an equivalence relation on E as follows: for e_1 and $e_2 \in E$, $e_1 \sim e_2$ if and only if $e_1 = e_2$ or $e_1 \neq e_2$ and there exists a circuit C containing both e_1 and e_2. The equivalence relation can be used to separate E into disjoint components. In each component every pair of elements is contained in a common circuit. If there is only one such component, then $\mathcal{E}$ is nonseparable; otherwise $\mathcal{E}$ is separable. For proofs of these matters, see [5].

In the following discussion we will be interested in finding the "least possible structure" needed on a set E to form an independence structure. One way of doing this is to consider the number of bases needed to insure that the ensuing structure $\mathcal{E}$ will satisfy (E_1), (E_2), and (E_3). If $|\mathcal{B}|$ is minimal for a fixed set E and given rank r, then the structure on E will be "small". If $\mathcal{E}$ is allowed to be separable, then it is easy to see that the smallest possible number for $|\mathcal{B}|$ is 1. Here $\mathcal{E}$ consists of r rank 1 components of 1 element each and one component of n - r elements of rank 0. The case in which $(E, \mathcal{E})$ is nonseparable is more difficult and is the one discussed below. Before proving the main theorem, we will need some further notation.

2. Restrictions and Contractions of Matroids

Given an independence structure $\mathcal{E}$ on E, it is possible to define several independent structures on subsets of E. Below we consider three types. The first is the dual $\mathcal{E}^*$ of $\mathcal{E}$ on E. $\mathcal{E}^* = \{A \subseteq E : A \subseteq E \setminus B, \text{ for some } B \in \mathcal{B}\}$. The bases $\mathcal{B}^*$ of $\mathcal{E}^*$ are just the complements, in E, of the bases of $\mathcal{E}$. We will use $\mathcal{C}^*$ and r^* to denote the circuits and rank function of $\mathcal{E}^*$ respectively.

If $|E| = n$, $r(E) = r$ then $r^*(E) = n - r$.

If $A \subseteq E$, then $\mathcal{E}_A$, the <u>restriction</u> of $\mathcal{E}$ to the set A, is defined by $\mathcal{E}_A = \{H \subseteq A : H \in \mathcal{E}\}$. We will use $\mathcal{C}_A$, r_A, and $\mathcal{B}_A$ to denote, respectively, the circuits, rank function, and bases of $\mathcal{E}_A$. Note that $\mathcal{C}_A \subseteq \mathcal{C}$ and that $\mathcal{C}_A$ consists of just those elements of $\mathcal{C}$ lying entirely within A.

If $A \subseteq E$, then $(\mathcal{E}^*)_A$ will denote the matroid $\mathcal{E}^*$ restricted to the subset A of E. We can carry the notation one step further: $((\mathcal{E}^*)_A)^*$. This is the dual, in A, of the matroid $(\mathcal{E}^*)_A$. This matroid is called the <u>contraction</u> of $\mathcal{E}$ to A and we will shorten the notation by writing $\mathcal{E}_{\otimes A}$ for $((\mathcal{E}^*)_A)^*$. We will write $\mathcal{C}_{\otimes A}$, $r_{\otimes A}$, and $\mathcal{B}_{\otimes A}$, respectively, for the circuits, rank function, and bases of $\mathcal{E}_{\otimes A}$.

Note that this definition of the contraction is equivalent to the other definitions in use [2, 3, 4]. For example, if we define $\mathcal{H} = \{D \subseteq A : D = C \cap A \neq \emptyset \text{ for } C \in \mathcal{C}\}$, then $\mathcal{C}_{\otimes A}$ consists of the minimal members of $\mathcal{H}$. Or, we can define $\mathcal{B}_{\otimes A} = \{H \subseteq A :$ there exists $B' \in \mathcal{B}_{E \setminus A}$ and $H \cup B' \in \mathcal{B}\}$. We leave it to the reader, or refer him to [3, 4] to show the equivalency of the definitions. In any case, the notation leads to the following proposition:

<u>Proposition 2.1. Let</u> $(E, \mathcal{E})$ <u>be a matroid and</u> $A \subseteq E$, <u>then</u>

(i) $(\mathcal{E}_{\otimes A})^* = (\mathcal{E}^*)_A$ and

(ii) $(\mathcal{E}_A)^* = (\mathcal{E}^*)_{\otimes A}$.

The key proposition needed to find the minimal nonseparable matroid is the following:

<u>Proposition 2.2.</u> <u>If</u> $(E, \mathcal{E})$ <u>is a nonseparable independence structure and</u> $e \in E$, $E' = E \setminus e$ <u>with</u> $\mathcal{E}_{E'}$ <u>separable, then</u> $\mathcal{E}_{\otimes E'}$ <u>is nonseparable.</u>

<u>Proof:</u> Let a and b be elements of different components of $\mathcal{E}_{E'}$. Since $\mathcal{E}$ is nonseparable, there is a circuit $C \in \mathcal{C}$ such that $a, b \in C$. Since a and b are in different components of $\mathcal{E}_{E'}$, $e \in C$ since $\mathcal{C}_{E'} = \{C \in \mathcal{C} : e \notin C\}$. Hence a and b are in $C \setminus e$ and $C \setminus e \in \mathcal{C}_{\otimes E'}$.

Now suppose a and b are in the same component of $\mathcal{E}_{E'}$. Since $\mathcal{E}_{E'}$ is separable, there is a $d \in E \setminus e$ in another component of $\mathcal{E}_{E'}$; hence using the argument above on the pairs of elements (a, d) and (b, d), there exist circuits C_a and C_b in $\mathcal{C}_{\otimes E'}$ such that $a, d \in C_a$ and $b, d \in C_b$. If $C_a = C_b$ we are done. If not, then since $d \in C_a \cap C_b$, we can apply Theorem 19 of [5] to deduce the existence of a circuit $C \in \mathcal{C}_{\otimes E'}$ containing both a and b. Hence every pair of elements of $(E', \mathcal{E}_{\otimes E'})$ lie in a common circuit and so $\mathcal{E}_{\otimes E'}$ is nonseparable. ▌

Note that H. H. Crapo proves a similar theorem with more complicated arguments in [2].

We shall also use the following in the proof of the main theorem:

if $(E, \mathcal{E})$ is a matroid, $x \in E$, $E' = E \setminus x$, then $|\mathcal{B}| = |\mathcal{B}_{E'}| + |\mathcal{B}_{\otimes E'}|$. The result follows since the first summand counts all those bases not containing x while the second counts those bases containing x. Finally, we use the facts, see [5], that the dual of a matroid is separable if and only if the matroid is separable and that the number of bases in a separable matroid is the product of the numbers of bases of the matroids formed by restricting the original one to each component.

3. The Minimal Nonseparable Matroid

The idea of the proof of the main theorem is to suppose that $(E, \mathcal{E})$ is nonseparable, $x \in E$, $E' = E \setminus x$. Then either $\mathcal{E}_{E'}$ or $\mathcal{E}_{\otimes E'}$ is nonseparable. We apply an appropriate induction hypothesis to the nonseparable one, approximate the number of bases in the other and use the fact that $|\mathcal{B}| = |\mathcal{B}_{E'}| + |\mathcal{B}_{\otimes E'}|$. To do this, we need the following counting lemma:

<u>Lemma 3.1.</u> <u>Let</u> $r, n \in \mathbb{Z}^+$ <u>with</u> $1 \leq r \lneq n$. <u>Suppose</u> $n_i, r_i \in \mathbb{Z}^+$ <u>for</u> $i = 1, 2, \ldots t$ <u>are given such that</u> $1 \leq r_i \leq n_i$ <u>and</u> $\sum_{i=1}^{t} r_i = r$ <u>and</u> $\sum_{i=1}^{t} n_i = n - 1$. <u>Then</u> $\prod_{i=1}^{t} [r_i(n_i - r_i) + 1] \geq n - r$ <u>for any</u> $t = 1, 2, \ldots, r$.

<u>Proof:</u> Let $k_i = n_i - r_i$, $i = 1, 2, \ldots, t$. Then $n - r = 1 + \sum_{i=1}^{t} (n_i - r_i)$ $= 1 + \sum_{i=1}^{t} k_i$, hence $\prod_{i=1}^{t} [r_i(n_i - r_i) + 1] = \prod_{i=1}^{t} (r_i k_i + 1)$ $\geq \sum_{i=1}^{t} r_i k_i + 1 \geq \sum_{i=1}^{t} k_i + 1 = n - r$, since $r_i \geq 1$ and $k_i \geq 0$ for each $i = 1, 2, \ldots, t$.

We now have enough machinery to prove

Theorem 3.2. Let $(E, \mathcal{E})$ be a nonseparable matroid with $|E| = n$ and $r(E) = r$. Then $|\mathcal{B}| \geq r(n-r)+1$. For any positive integers n and r with $n > r$, there exists one (up to an isomorphism) nonseparable matroid $(E, \mathcal{E})$ with $|E| = n$, $r(E) = r$ and $|\mathcal{B}| = r(n-r)+1$.

Proof: The proof is by induction on n and r. The cases $n = 1$ and $r = 1$ and $r = 1$, $n \geq 2$ are trivially true. For $n = r+1$, there is only one (up to an isomorphism) nonseparable matroid since the dual of such a structure is a rank 1 nonseparable matroid. Hence the theorem is true for $r+1 = n$.

We now assume that the theorem holds for all non-separable independence structures of rank less than or equal to a given r on sets of cardinality less than n. We let $(E, \mathcal{E})$ be a nonseparable matroid with $|E| = n$, $r(E) = r$ and let $x \in E$, $E' = E \setminus x$. Then one of the following four cases must be satisfied: (i) both $\mathcal{E}_{E'}$ and $\mathcal{E}_{\otimes E'}$ are nonseparable; (ii) $\mathcal{E}_{E'}$ is separable, but $\mathcal{E}_{\otimes E'}$ is nonseparable; (iii) $\mathcal{E}_{E'}$ is non-separable and $\mathcal{E}_{\otimes E'}$ is separable; (iv) both $\mathcal{E}_{\otimes E'}$ and $\mathcal{E}_{E'}$ are separable.

Case (iv) does not occur since by proposition 2.2, if $\mathcal{E}_{E'}$ is separable, $\mathcal{E}_{\otimes E'}$ is nonseparable. Using duality and proposition 2.1, we can reduce case (iii) to case (ii).

Now if either case (i) or (ii) occurs, then $\mathcal{E}_{\otimes E'}$ is nonseparable of rank $r_{\otimes E'}(E') = r - 1$ since x must lie in some basis of $\mathcal{E}$, hence, by the induction hypothesis applied to $(E', \mathcal{E}_{\otimes E'})$ we have that $|\mathcal{B}_{\otimes E'}| \geq (r-1)(n-r) + 1$. Since $\mathcal{E}$ is nonseparable, $r(E') = r$ since x is not in every basis of $\mathcal{E}$. We suppose that $\mathcal{E}_{E'}$ has t $(t = 1, 2, \ldots, r)$ components $E_1, E_2, \ldots, E_t$ such that $|E_i| = n_i$, $r_{E'}(E_i) = r_i$, then we have $\sum_{i=1}^{t} n_i = n - 1$, $\sum_{i=1}^{t} r_i = r$ and since each component is nonseparable $n_i \geq r_i$, $i = 1, 2, \ldots, t$. We deduce from the induction hypothesis (including case (i), $t = 1$) that $|\mathcal{B}_{E_i}| \geq r_i(n_i - r_i) + 1$. From lemma 3.1, we see that in both cases (i) and (ii), $|\mathcal{B}_{E'}| \geq \prod_{i=1}^{t} [r_i(n_i - r_i) + 1] \geq n - r$. Hence, if $\mathcal{E}_{E'}$ is separable or not $|\mathcal{B}| = |\mathcal{B}_{\otimes E'}| + |\mathcal{B}_{E'}| \geq (r-1)(n-r) + 1 + n - r$ $= r(n-r) + 1$ which proves the first part of the theorem.

Note that the only case where equality can occur, i.e., $|\mathcal{B}| = r(n-r) + 1$, happens if both $|\mathcal{B}_{\otimes E'}| = (r-1)(n-r) + 1$ and $|\mathcal{B}_{E'}| = n - r$. The first occurs if $\mathcal{E}_{\otimes E'}$ is the unique minimal structure (by the induction hypothesis) while the second occurs if $\mathcal{E}_{E'}$ consists of r rank 1 components, $r - 1$ of them containing one element, the last containing $n - r$ elements. These two occur together if and only if the set $E = \{e_1, \ldots, e_n\}$ has circuit structure

$\mathcal{C} = \{\{e_1, e_2, \ldots, e_r, e_j\} \{e_i, e_j\}_{i \neq j} : i, j = r+1, r+2, \ldots, n\}$. ∎

We will call a nonseparable matroid of rank r on n elements with $|\mathcal{B}| = r(n-r) + 1$ a <u>minimal matroid</u>.

4. The "Second Minimal" Nonseparable Matroid

If $(E, \mathcal{E})$ is a nonseparable matroid with $|E| = n$, $r(E) = r$ then we know that $|\mathcal{B}| \geq r(n-r)+1$. It is not, in general, possible that $|\mathcal{B}| = r(n-r)+2$ and in fact if $\mathcal{E}$ is not precisely the minimal structure we have defined above, we shall show that if $2r \leq n$, $|\mathcal{B}| \geq 2(r-1)(n-r-1)+1+r$ and if $2r \geq n$ then $|\mathcal{B}| \geq (2r-1)(n-r-1)+2$. In each case, except for $2r = n$, there is precisely one, up to an isomorphism, nonseparable matroid where equality occurs in the above equations. When $2r = n$, there are two such matroids which are duals of each other.

To prove this result, we will use the following sequence of lemmas and throughout the sequel we will assume that $(E, \mathcal{E})$ is a nonseparable matroid with $|E| = n$, $r(E) = r$, $x \in E$, $E' = E \setminus x$ and $E = \{e_1, e_2, \ldots, e_n\}$.

<u>Lemma 4.1.</u> <u>If</u> $\mathcal{E}_{\otimes E'}$ <u>is separable and</u> $C \in \mathcal{C}_{\otimes E'}$, <u>then either</u> $C \in \mathcal{C}_{E'}$ <u>or there exists</u> $C' \in \mathcal{C}_{\otimes E'} \setminus \mathcal{C}_{E'}$ <u>such that</u> $C \cup C' \in \mathcal{C}_E$.

<u>Proof:</u> Suppose $C \in \mathcal{C}_{\otimes E'} \setminus \mathcal{C}_{E'}$, then $C \cup x \in \mathcal{C}$. Since $\mathcal{C}_{\otimes E'}$ is separable, there is a $y \in E'$ such that C and y lie in different components of $\mathcal{C}_{\otimes E'}$. Since $\mathcal{C}$ is nonseparable, there exists $D \in \mathcal{C}$ containing both x and y, hence $D' = D \setminus x \in \mathcal{C}_{\otimes E'}$ and $D \cap C = \emptyset$ since C and y lie in different components of $\mathcal{C}_{\otimes E'}$.

From the circuit axiom (C_2) in Sec. 1, we can deduce the existence of $F \in \mathcal{C}$ such that $F \subseteq [(C \cup x) \cup D] \setminus x$ and suppose that

$F \subsetneqq C \cup D'$ otherwise the theorem obtains. We can suppose then that $C \setminus F \neq \emptyset$ and we know $D' \cap F \neq \emptyset$ otherwise $F \subsetneqq C \cup x$ or D. So let $c \in C \setminus F$ and $f \in D' \cap F$. Using axiom (C_2) again, we can find $G \in \mathcal{C}$ such that $x \in G \subseteq (D \cup F) \setminus f$ and hence $G' = G \setminus x \in \mathcal{C}_{\otimes E'}$. Since $c \in C$, $f \in D' \setminus G'$, then $G' \cap C \neq \emptyset$ and $G' \cap D' \neq \emptyset$ otherwise G is a proper subset of either C or D; but then $G' \in \mathcal{C}_{\otimes E'}$ connects two different components of $\mathcal{C}_{\otimes E'}$ which can't happen. Hence our assumption that $C \setminus F \neq \emptyset$ is false; similarly $D' \setminus F \neq \emptyset$ can't happen, so that $F = D' \cup C$. ▌

<u>Corollary 4.2. If</u> $C_1, C_2 \in \mathcal{C}_{\otimes E'} \setminus \mathcal{C}_{E'}$ <u>lie in different components of</u> $\mathcal{C}_{\otimes E'}$, <u>then</u> $C_1 \cup C_2 \in \mathcal{C}_{E'}$.

<u>Corollary 4.3. Each element of</u> E' <u>which is not a dependent singleton of</u> $\mathcal{E}_{\otimes E'}$ <u>lies in a component of</u> $\mathcal{E}_{\otimes E'}$ <u>containing at least 2 elements.</u>

<u>Lemma 4.4. Suppose</u> $\mathcal{C}_{\otimes E'} = \{\{e_1, \ldots, e_r\}, \{e_j\} : j = r+1, \ldots, n-1\}$ $(x = e_n)$; then $\mathcal{E}$ must be a minimal matroid.

<u>Proof:</u> Lemma 4.1 insures that $\mathcal{C} \supseteq \{\{x, e_j\}, \{e_1, e_2, \ldots, e_r, x\}, \{e_1, e_2, \ldots, e_r, e_j\}, \{e_i, e_j\}_{i \neq j} : i, j = r+1, \ldots, n-1$ which is a non-separable minimal matroid. Using axioms (C_1) and (C_2) of Sec. 1, one can easily show that it is not possible to add further circuits to $\mathcal{C}$ and keep $\mathcal{C}_{\otimes E'}$ as described. ▌

The two matroids we define below are the two "second minimal" nonseparable matroids on a set of n elements with rank r. In the theorem we will prove, we show that these two have the next largest number of bases possible after the minimal structure.

Definition: If $(E, \mathcal{E})$ is a matroid with $\mathcal{C} = \{\{e_1, e_2, \ldots, e_{r+1}\}, \{e_1, e_2, \ldots, e_{r-1}, e_j\}, \{e_r, e_{r+1}, e_j\}, \{e_i, e_j\}_{i \neq j} : i, j = r+2, \ldots, n\}$ then $\mathcal{E}$ is a nonseparable matroid of rank r on n elements. We will call such a structure a form I(n, r) matroid or an I(n, r) matroid. The dual of a I(n, r) matroid is of rank n − r on n elements. We will call the dual of a I(n, r) matroid a II(n, n−r) matroid.

The circuits of a II(n, n−r) matroid are easily seen to be $\mathcal{C} = \{\{e_r, e_{r+1}\}, \{e_i, e_j\}_{i \neq j}, \{e_i, e_k, e_{r+2}, \ldots, e_n\} : i, j = 1, 2, \ldots, r-1,$ and $k = r, r+1\}$. If $(E, \mathcal{E})$ is a I(n, r) matroid, $|\mathcal{B}| = 2(r-1)(n-r-1) + 1 + r$ while if it is a II(n, r) matroid (the dual of a I(n, n−r) structure), $|\mathcal{B}| = (2r-1)(n-r-1) + 2$.

Lemma 4.5. Suppose $(E, \mathcal{E})$ is not minimal but $\mathcal{E}_{E'}$ is a minimal matroid and $\mathcal{E}_{\otimes E'}$ is separable, then $\mathcal{E}_{\otimes E'}$ is separable into two components, one of rank $i-1$ on i elements, the other a minimal structure of rank $r-i$ on $n-i-1$ elements and $|\mathcal{B}_{\otimes B'}| = i(r-i)(n-r-1) + i$ for $1 \leq i \leq r-1$.

Proof: Note that $\mathcal{C}_{E'} \supsetneqq \mathcal{C}_{\otimes E'}$. Suppose, after a suitable reordering of the elements of E, that $\mathcal{C}_{E'} = \{\{e_1, e_2, \ldots, e_{r-1}, e_r, e_i\},$

$\{e_i, e_j\}_{i \neq j} : i, j \in \{r+1, r+2, \ldots, n-1\} =: J\}$ and $x = e_n$. Since $\mathcal{E}_{\otimes E'}$ is separable, there exists $D \in \mathcal{C}_{\otimes E'}$ with $D \cup x \in \mathcal{C}$. Note $D \neq \emptyset$ and $D \neq e_j$, $j \in J$ for if that were so then $\{x, e_j\} \in \mathcal{C}$ for $j \in J$ and hence $\mathcal{E}$ would be minimal. Let $C_j = \{e_1, e_2, \ldots, e_r, e_j\}$, $j \in J$. Then $D \subseteq C_j$ for some $j \in J$, otherwise $D \cup x$ would properly contain a circuit of $\mathcal{E}$ contradicting the minimality of circuits. Since $\mathcal{E}_{\otimes E'}$ is separable, we know, from lemma 4.1, there exists a nonempty set $H \in \mathcal{C}_{\otimes E'}$ such that $H \cup x \in \mathcal{C}$ and $D \cup H \in \mathcal{C}_{E'}$. The only possible choice for H is $H = C_j \setminus D$ for some $j \in J$.

Either H or D contains e_j; without loss of generality we assume $e_j \in D$. Then, since axiom (C_2) shows $(x \cup D \cup \{e_j, e_k\}) \setminus e_j$ contains a circuit for each $k \in J$, $k \neq j$, we see that in fact $(x \cup D \cup e_k) \setminus e_j \in \mathcal{C}$ and hence $(D \cup e_k) \setminus e_j \in \mathcal{C}_{\otimes E'}$ for each $k \in J$. And so each of H, $(D \cup e_k) \setminus e_j$, and $\{e_j, e_k\}$ $j \neq k$, $j, k \in J$ are in $\mathcal{C}_{\otimes E'}$. Since addition of further circuits to $\mathcal{C}_{\otimes E'}$ would contradict the circuit axioms or add circuits to $\mathcal{C}_{E'}$, these are the only circuits in $\mathcal{C}_{\otimes E'}$ and hence the matroid $\mathcal{C}_{\otimes E'}$ has the desired structure.

To find $|\mathcal{B}_{\otimes E'}|$, let $|H| = i$, then $r_{\otimes E'}(H) = i - 1$, $r_{\otimes E'}(E' \setminus H) = (r-1) - (i-1) = r - i$, and $|E' \setminus H| = n - i - 1$. Since the two components of $\mathcal{E}_{\otimes E'}$ are H and $E' \setminus H$, then $|\mathcal{B}_{\otimes E'}| = |(\mathcal{B}_{\otimes E'})_H| \cdot |(\mathcal{B}_{\otimes E'})_{E' \setminus H}| = i[(r-i)(n-r-1)+1]$ since $E' \setminus H$ is a minimal matroid. Note $1 \leq i \leq r-2$ since both $H \neq \emptyset \neq D$. ▌

Remark 4.6. If $(E, \mathcal{E})$ satisfies the conditions of the lemma, then $|\mathcal{B}| = [r(i+1) - 1](n - r - 1) + 1 + i$.

Remark 4.7. If $(E, \mathcal{E})$ satisfies the conditions of the lemma and $|H| = 1$, then $\mathcal{E}$ is a $\mathrm{II}(n, r)$ matroid.

Lemma 4.8. Suppose $\mathcal{E}_{\otimes E'}$ is separable, consisting of 2 components—one of rank one on 2 elements, the other of rank $r - 2$ on $r - 1$ elements with the remainder dependent singletons. Then $(E, \mathcal{E})$ is a $I(n, r)$ matroid.

To prove this it suffices to show that after a suitable reordering of the elements of E, $\mathcal{C} \supseteq \{\{e_1, e_2, \dots, e_{r-1}, e_j\}, \{x, e_j\}, \{e_j, e_k\}_{j\neq k}, \{e_1, e_2, \dots, e_{r-1}, e_r, r_{r+1}\}, \{x, e_r, e_{r+1}\} : j, k \in \{r+2, \dots, n-1\}\}$ and that no other circuits are possible given this set. This is straightforward and the proof is left to the reader.

Lemma 4.9. Let $t, n, r_i, k_i \in \mathbf{Z}^+ \cup \{0\}$, $i = 1, 2, \dots, t$ satisfying $n > r > 1$, $r_i k_i \geq 1$, for $i = 1, 2, \dots, t$, $\sum_{i=1}^{t} r_i = r - 1$ and $\sum_{i=1}^{t} (r_i + k_i) = n - 1$. Then if $t \geq 3$, $\prod_{i=1}^{t} (r_i k_i + 1) \gneq 2r - 1$. If $t = 2$ and at least one of k_1 and $k_2 \geq 2$ or both $r_1 \geq 2$ and $r_2 \geq 2$, then $\prod_{i=1}^{2} (r_i k_i + 1) \geq 2r - 1$ with equality if and only if $r_1 = r_2 = 2$ and $k_1 = k_2 = 1$.

Proof: For $t \geq 3$, $\prod_{i=1}^{t} (r_i k_i + 1) = \prod_{i=1}^{t} r_i k_i + \dots + \sum_{1 \leq i \lneq j \leq t} (r_i k_i)(r_j k_j)$

$$+ \sum_{i=1}^{t} r_i k_i + 1 > (r_1k_1)(r_2k_2) + (r_2k_2)(r_3k_3) + \sum_{j=3}^{t} (r_1k_1)(r_jk_j) + \sum_{i=1}^{t} (r_ik_i) + 1$$

$$\geq 2 \sum_{i=1}^{t} r_i k_i + 1 \geq 2 \sum_{i=1}^{t} r_i + 1 = 2(r - 1) + 1 = 2r - 1$$ since $r_1 k_1 \geq 1$ and $\sum_{i=1}^{t} r_i = r - 1$.

If $t = 2$, $(r_1k_1 + 1)(r_2k_2 + 1) - (2r - 1) = (r_1k_1 + 1)(r_2k_2 + 1) - (2(r_1 + r_2 + 1) - 1) = (r_1(k_1 - 2) + 1)(r_2(k_2 - 2) + 1) + 2r_1r_2(k_1 + k_2 - 2) - 1 \geq 0$ as long as either $k_1 \geq 2$ or $k_2 \geq 2$ or both $r_1 \geq 2$ and $r_2 \geq 2$ with equality if and only if $r_1 = r_2 = 2$ and $k_1 = k_2 = 1$. ∎

We can now prove

Theorem 4.10. Let (E, ξ) be a nonseparable matroid with $|E| = n$, $r(E) = r$. Then either ξ is a minimal matroid; or, if $2r \geq n$, ξ is a $I(n, r)$ matroid; or contains more bases than a $I(n, r)$ matroid, or if $2r \geq n$, ξ is a $II(n, r)$ matroid, or ξ contains more bases than a $II(n, r)$ matroid.

Proof: We let $x \in E$, $E' = E \setminus x$, then as in theorem 3.2, we need only consider 3 cases, namely: (i) both $\xi_{E'}$ and $\xi_{\otimes E'}$ are nonseparable; (ii) $\xi_{E'}$ is nonseparable, but $\xi_{\otimes E'}$ is separable; and (iii) $\xi_{E'}$ is seprable and $\xi_{\otimes E'}$ is nonseparable.

If case (i) holds, then since both $\xi_{E'}$ and $\xi_{\otimes E'}$ are nonseparable, then both must contain as many bases as a minimal matroid. Hence

$|\mathcal{B}_{E'}| \geq r(n - r - 1) + 1$ and $|\mathcal{B}_{\otimes E'}| \geq (r - 1)(n - r) + 1$ so that $|\mathcal{B}| = |\mathcal{B}_{E'}| + |\mathcal{B}_{\otimes E'}| \geq (2r - 1)(n - r - 1) + 1 + r > (2r - 1)(n - r - 1) + 2$ if $r > 1$. If $r = 1$, ξ must be a minimal structure. Since $(2r - 1)(n - r - 1) + 2 \geq 2(r - 1)(n - r - 1) + 1 + r$ when $2r \leq n$, the theorem always holds in this case for then ξ contains more bases than either a $I(n, r)$ or $II(n, r)$ matroid.

To prove the theorem when case (ii) obtains we follow the same sort of procedure as in theorem 3.2, we use induction on n for a fixed r. The first nontrivial case occurs with $r(E) = r$, $|E| = r + 1$. Then, from theorem 3.2, ξ must be a minimal structure and the theorem obtains.

We now assume the theorem holds for all nonseparable matroids on rank r or less on fewer than n elements and we assume that $|E| = n$, $r(E) = r$, $\xi_{E'}$ nonseparable, and $\xi_{\otimes E'}$ separable. We break this case up into several subcases: (a) $\xi_{E'}$ is minimal, (b) $\xi_{E'}$ is not minimal and $\xi_{\otimes E'}$ contains precisely one nontrivial component, (c) $\xi_{E'}$ is not minimal and $\xi_{\otimes E'}$ contains two nontrivial components, and (d) $\xi_{E'}$ is not minimal and $\xi_{\otimes E'}$ contains 3 or more nontrivial components.

If subcase (a) obtains, from remark 4.6, we see that $|\mathcal{B}| = (r(i+1) - 1)(n-r-1) + 1 + i \geq (2r-1)(n-r-1) + 2$ for $1 \leq i \leq r - 2$, with equality if and only if $i = 1$. When $i = 1$, we see from remark 4.9 that ξ must be a $II(n, r)$ matroid. Since $(2r-1)(n-r-1) + 2 \geq 2(r-1)(n-r-1) + 1 + r$ when $2r \leq n$ and the number of bases in a $I(n, r)$

matroid is $2(r-1)(n-r-1) + 1 + r$, we see that either $\mathcal{E}$ is a $\mathrm{II}(n, r)$ matroid or contains more bases than a $\mathrm{II}(n, r)$ matroid and hence the theorem obtains.

In subcases (b), (c) (and d) we have that $|\mathcal{E}_{E'}| \geq (2r-1)(n-r-2) + 2$ if $2r \geq n - 1$ and $|\mathcal{B}_{E'}| \geq 2(r-1)(n-r-1) + 1 + r$ if $2r \leq n - 1$. If we show that in either case $|\mathcal{B}_{\otimes E'}| \geq 2r - 1$, these subcases will also obtain.

In subcase (b), $\mathcal{E}_{\otimes E'}$ contains one nontrivial component, it may contain i other dependent singletons. Hence $|\mathcal{B}_{\otimes E'}| \geq (r-1)(n-i-1-(r-1)) + 1$ since the nontrivial component must contain at least as many bases as a minimal matroid. Note that $n - r - i \geq 2$ otherwise the nontrivial component is of rank $r - 1$ in $\mathcal{E}_{\otimes E'}$, on r elements and hence by lemma 4.4, $\mathcal{E}$ is a minimal structure; hence if $\mathcal{E}_{E'}$ is nonseparable, it must be minimal. If $n - r - i = 2$ and the nontrivial component of $\mathcal{E}_{\otimes E'}$ is a minimal matroid, it is of rank $r - 1$ on $r + 1$ elements, hence $|\mathcal{B}_{\otimes E'}| = 2r - 1$. If $\mathcal{E}_{E'}$ is precisely a $\mathrm{II}(n-1, r)$ matroid, then it can be easily seen that then $\mathcal{E}$ must be a $\mathrm{II}(n, r)$ matroid whence the theorem holds.

If $\mathcal{E}_{E'}$ is not a form $\mathrm{II}(n-1, r)$ matroid or if $n - r - i \geq 3$, then the theorem follows immediately; for if the former holds, then either $|\mathcal{B}_{E'}| > (2r-1)(n-r-2) + 2$ or $|\mathcal{B}_{E'}| \geq 2(r-1)(n-r-2) + 1 + r$, and in both these cases, adding $2r - 1$ bases yields more bases than either a $\mathrm{I}(n, r)$ or $\mathrm{II}(n, r)$ matroid. If the latter case holds, we have the same situation,

$|\mathcal{B}_{\otimes E'}| \geq 3r - 2$ which when added to $|\mathcal{B}_{E'}|$ yields too many bases for either a $I(n, r)$ or $II(n, r)$ matroid.

To prove the theorem under the hypothesis of subcase (c), let r_1 and r_2 be the ranks in $\mathcal{E}_{\otimes E}$ of the two nontrivial components E_1 and E_2 of $\mathcal{E}_{\otimes E'}$. Let k_1 and k_2 be chosen so that $k_1 + r_1 = |E_1|$ and $k_2 + r_2 = |E_2|$. Note that $r_1 + r_2 = r - 1$, but that $r_1 + r_2 + k_1 + k_2 \leq n - 1$ since there may be dependent singletons in $\mathcal{E}_{\otimes E'}$. If $k_1 = k_2 = 1$ and $r_1 = 1$, then the conditions of lemma 4.8 are satisfied and hence $\mathcal{E}$ is a $I(n, r)$ matroid. If $k_1 = k_2 = 1$ and $r_1, r_2 \geqq 2$, or if $k_1 \geqq 2$ or $k_2 \geqq 2$ then the conditions of lemma 4.9 are satisfied. Each component of $\mathcal{E}_{\otimes E'}$ is at least a minimal matroid, hence $|\mathcal{B}_{\otimes E'}| \geq (r_1 k_1 + 1)(r_2 k_2 + 1) > 2r - 1$, and using the same arguments as above we see that $\mathcal{E}$ contains more bases than a $I(n, r)$ matroid if $2r \leq n$ or a $II(n, r)$ matroid if $2r \geqq n$.

In the case $k_1 = 1 = k_2$ and $r_1 = r_2 = 2$, then $\mathcal{E}$ is of rank 5 on at least 7 elements. It is easily seen that $\mathcal{E}_{E'}$ cannot be a minimal nor $I(n-1, r)$ nor $II(n-1, r)$ matroid and hence the number of bases in $\mathcal{E}_{E'}$ is too large for $\mathcal{E}$ to be a $I(n, r)$ matroid if $2r \leq n$ or a $II(n, r)$ matroid if $2r \geq n$. In any case, the theorem obtains.

If case (d) obtains, we let $r_1, r_2, \ldots, r_t$ be the ranks in $\mathcal{E}_{\otimes E'}$ of the $t \geqq 3$ nontrivial components $E_1, E_2, \ldots, E_t$ of $\mathcal{E}_{\otimes E'}$ and suppose $k_1, k_2, \ldots, k_t$ are chosen so that $r_1 + k_1 = |E_1|, \ldots, r_t + k_t = |E_t|$.

Then $k_i \geq 1$ for each $i \leq t$ and the conditions of lemma 4.9 are satisfied. Hence, since each component must contain at least as many bases as a minimal matroid, $|(\mathcal{B}_{\otimes E'})_{E_i}| \geq r_i k_i + 1$, hence

$|\mathcal{B}_{\otimes E'}| \geq \prod_{i=1}^{t} (r_i k_i + 1) > 2r - 1$ since $t \geq 3$. Whence as in part (c) the theorem obtains. This completes the proof of case (ii).

If case (iii) obtains, we need only look at the dual structure $\mathcal{E}^*$, $(\mathcal{E}_{E'})^*$, and $(\mathcal{E}_{\otimes E'})^*$. Using proposition 2.1, this reduces case (iii) to case (ii) above on the dual. Since I(n, r) and II(n-r, r) matroids are duals of each other and since $2r \leq n$ implies that $2(n-r) \geq n$ and conversely, we see that the theorem obtains for the case also.

Bibliography

[1] D. S. Asche, Minimal Dependent Sets, J. of the Australian Math. Soc., vol. 6 (1966), 259-262.

[2] H. H. Crapo, A Higher Invariant for Matroids, J. of Combinatorial Th. vol. 2 (1967), 406-417.

[3] H. H. Crapo and G. C. Rota, On the Foundations of Combinatorial Geometries, December 1968. Privately published.

[4] W. T. Tutte, Lectures on Matroids, J. Res. Nat. Bur. Standards, Sec. B 69 (1965), 1-53.

[5] Hassler Whitney, On the abstract properties of linear independence, Amer. J. Math. vol. 57 (1935), 509-533.

University of Wisconsin, Madoson, Wisconsin, U. S. A.

REPRESENTATION SUR UN CORPS DES MATROIDES D'ORDRE ≤ 8

J. C. Fournier

Résumé: On démontre que tout matroïde d'ordre 7 (i.e. sur un ensemble de cardinal 7) et d'ordre 8 de rang ≠ 4 est rèprésentable sur un corps commutatif.

Introduction. Lazarson[1] donna par des considérations de caractéristiques des corps un exemple de matroïde d'ordre 16, non représentable sur un corps. Ingleton [2] trouva ensuite par des considérations géométriques un exemple de nature différente et plus simple puisque d'ordre 9 : le matroïde de rang 3 défini dans le plan projectif $P_2(\mathbb{R})$

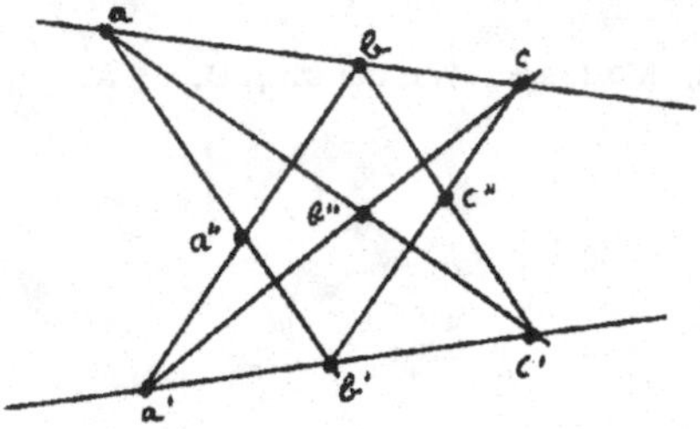

par la configuration de Pappus à laquelle on enlève une droite, a"b"c" sur la figure ci-contre. Suivant le théorème de Pappus, dans toute représentation de ce matroïde sur un corps les points a",b",c", sont alignés et par conséquent une telle représentation ne peut exister.

Ingleton conjecture que cet exemple (et son dual) est le plus petit. Il â déja été démontré que tous les matroïdes d'ordre ≤ 6 sont représentables sur $\mathbb{Q}$. On connait en outre un matroïde d'ordre 7 de rang 3 (et son dual qui est de rang 4) représentable uniquement sur CG(2) (corps de Galois à deux éléménts) ou une extension de ce corps :

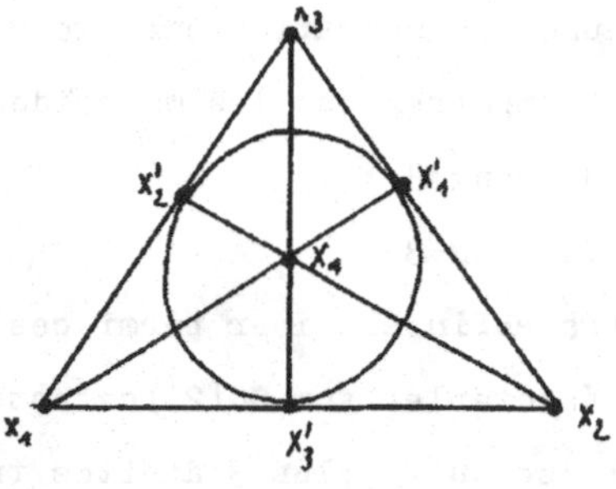

c'est celui défini par la configuration de Fano; c'est encore $P_2[CG(2)]$ i.e. le plan projectif construit sur CG(2). Dans toute représentation x'_3 par exemple est conjugué harmonique de lui-même par rapport à x_1 et x_2, le corps est donc nécessairement de caractéristique 2;

Depuis Vámos a trouvé un matroïde d'ordre 8 et de rang 4 non représentable et d'autres ensuite furent trouvés d'ordre 8 mais tous de rang 4.

Préliminaires. Tout matroïde $\mathcal{M}(E)$ considéré ici est géométrique i.e. vérifie $\bar{\emptyset}=\emptyset$ et $\overline{\{x\}}=\{x\}$ pour tout $x\in E$, hypothèse qui ne restreint aucunement la généralité des résultats (en considérant pour un matroïde quelconque le matroïde géométrique associé).

On se propose de montrer que les matroïdes :

- d'ordre 7
- d'ordre 8 et de rang $\neq 4$

sont représentables sur un corps commutatif.

Observons d'abord que

- tout matroïde de rang 2 est trivialement représentable.
- si un matroïde est représentable, son dual l'est aussi.

(sur le même corps -cf pour une démonstration par exemple[3]).

Il en résulte qu'il suffit de montrer que les matroïdes

- d'ordre 7 et de rang 3
- d'ordre 8 et de rang 3

sont représentables. On pourra enfin éliminer parmi ces matroïdes de rang 3 les binaires (représentables sur CG(2)) caractérisés par le fait que par tout point ne passe qu'au plus 3 droites triviales ou non. (Dans toute la suite un fermé de rang 2 ayant au moins 3 points est une <u>droite</u> et une <u>droite triviale</u> s'il n'a que 2 points).

I- <u>Tout matroïde d'ordre 7 est représentable sur CG(2) ou $\mathbb{Q}$</u>

<u>Lemme 1</u>: <u>Le matroïde de Fano est le seul de rang 3 ayant 7 points et au moins 7 droites</u> (cf [3] pour démonstration de ce lemme trés simple)

Soit donc $\mathcal{M}(E)$ de rang 3, $|E| = 7$. Excluant les matroïdes binaires, on peut supposer qu'il existe un point $e \in E$ par lequel passe au moins 4 droites triviales ou non : ex_1, ex_2, ex_3, ex_4; comme : $|E - \{e, x_1, x_2, x_3, x_4\}| = 2$ il ne passe par e qu'au plus 2 droites (non triviales).

<u>Ier Cas</u>. Il existe un point par lequel ne passe aucune droite. Soit s une représentation (s est une injection de E-e dans $P_2(\mathbb{Q})$) dans $P_2(\mathbb{Q})$ du sous-matroide à 6 points $\mathcal{M}(E-e)$ et e' un point de $P_2(\mathbb{Q})$ n'appartenant à aucune des droites de $P_2(\mathbb{Q})$ définies par l'image par s de 2 points de E-e. En prenant e' comme image de e, la représentation de $\mathcal{M}(E-e)$ se prolonge en une représentation de $\mathcal{M}(E)$ sur $\mathbb{Q}$. ($\mathcal{M}(E)$ et le matroïde induit par $P_2(\mathbb{Q})$ sur s(E-e) +e' ont leurs droites en correspondance bijective par s).

<u>2ème Cas</u>. Il existe un point $e \in E$ par lequel ne passe qu'une seule droite d. Le sous matroïde à 6 points $\mathcal{M}(E-e)$ admet une représentation dans $P_2(\mathbb{Q})$. Soit d' la droite de $P_2(\mathbb{Q})$ définie par les images dans $P_2(\mathbb{Q})$ des points de d-e. Soit enfin e' un point de d' n'appartenant à aucune des droites de $P_2(\mathbb{Q})$ définies par la représentation de $\mathcal{M}(E-e)$ (e' existe car d' a une infinité de points alors que $\mathcal{M}(E-e)$ n'a qu'un nombre fini de droites triviales ou non). En prenant e' pour image de e on vérifie aisément que comme dans le premier cas la représentation de $\mathcal{M}(E-e)$ se prolonge en une représentation de $\mathcal{M}(E)$ sur $\mathbb{Q}$.
<u>3ème Cas</u>. Par tout point passe au moins 2 droites. Soit alors e le point par lequel ne passe que 2 droites $d_1 = \{e,a,b,\ldots\}$ $d_2 = \{e,c,d,\ldots\}$, $s : E-e \to P_2(\mathbb{Q})$ une représentation de $\mathcal{M}(E-e)$, e' le point d'intersection des 2 droites d_1' et d_2' définies dans $P_2(\mathbb{Q})$ respectivement par s(a),s(b) et s(c),s(d). (e' existe dans $P_2(\mathbb{Q})$ car d_1' et d_2' sont définies par des points à coordonnées rationnelles).
Supposons qu'il existe 2 points x' et y' distincts de a,b,c,d, alignés avec e' dans $P_2(\mathbb{Q})$ et considérons le matroïde induit par $P_2(\mathbb{Q})$ sur l'ensemble s(E-e) + e'; sur s(E-e) ce matroïde est isomorphe à $\mathcal{M}(E-e)$, il passe donc par chacun des points x et y au moins 2 droites toutes nécessairement distinctes de la droite {e',x',y'} (en effet cette droite n'est pas une droite de $\mathcal{M}(E)$ puisque par e ne passe dans $\mathcal{M}(E)$ que 2 droites). Avec les droites d_1' et d_2', le matroïde a donc au moins 7 droites ; d'après le lemme 1, il n'est pas représentable sur $\mathbb{Q}$. Par conséquent par e' ne passe pas d'autre droite définie dans $P_2(\mathbb{Q})$ par s(E-e) que d_1' et d_2' . Comme au 2ème cas, il en résulte qu'avec e' = s(e), on obtient une représentation de sur $\mathbb{Q}$, ce qui achève la démonstration de I.

II- <u>Tout matroïde d'ordre 8 et de rang 3 est représentable soit sur $\mathbb{Q}$ soit sur un corps fini de caractéristique 2</u>

<u>Lemme 2</u> : <u>Il n'existe qu'un matroïde d'ordre 8 de rang 3 non binaire et tel qu'en chaque point passe au moins 3 droites. Il est représentable sur une extension finie de CG(2) mais pas sur $\mathbb{Q}$.</u>

<u>Démonstration</u> : Soit $\mathcal{M}(E)$ vérifiant les hypothèses du lemme ; $\mathcal{M}$ n'étant pas binaire il existe $e \in E$ par lequel passe au moins 4 droites triviales ou non. Par e passe déjà 3 droites : $d_1 = \{e\ a\ a'\}$ $d_2 = \{e\ b\ b'\}$, $d_3 = \{e\ c\ c'\}$; la 4ème ne peut être que triviale (car $|E| = 8$) : $d_4 = \{e\ f\}$. Par f passe 3 droites d'_1 , d'_2 , d'_3 ,qui n'étant pas triviales ne contiennent pas e et sont donc chacune définies par 2 des 6 points : a,a', b,b', c,c'; 2 d'_i ne se coupant qu'en f, chacune d'elles ne contient que 2 de ces 6 points. On a donc nécessairement une configuration comme celle-ci :

$$d'_1 = \{f\ a\ b'\}\ ,\quad d'_2 = \{f\ b\ c'\}\ ,\quad d'_3 = \{f\ a'\ c\}\ .$$

En a doit passer une 3ème droite autre que d_1 et d'_1 ; comme seuls les points $b,c,c' \notin d_1 \cup d'_1$, celle-ci est nécessairement : $d = \{a\ b\ c\}$ (car les droites bc' et cc' ne contiennent pas a). <u>De même</u> enfin en a' passe une 3ème droite : $d' = \{a'\ b'\ c'\}$.

Il ne peut y avoir d'autres droites puisqu'en chaque point passe déjà 3 droites non triviales (une 4ème exigerait $2 \times 4 + 1 = 9$ points dans E, or $|E| = 8$).

Ce matroïde est ainsi complètement déterminé et donc unique en son genre ; en outre sa configuration est totalement symétrique par rapport aux sommets tout comme celle de Fano (comme on le verra par la suite, ce matroïde joue dans la classe des matroides d'ordre 8 un rôle analogue à celui de Fano pour l'ordre 7).

Etude de la représentation de ce matroïde $\mathcal{M}(E)$:

Soit donnée une représentation sur le corps K de ce matroide sous la forme : $E = \{a,b,c,a',b',c',e,f\} \subset P_2(K)$

Alors $\{e,a,b\}$ est une base de $P_2(K)$; posons

$a' = e + \lambda a, \quad b' = e + \mu b, \quad c' = e + \nu c, \quad c = a + \alpha b$

où $\lambda, \mu, \nu, \alpha \in K - \{0\}$

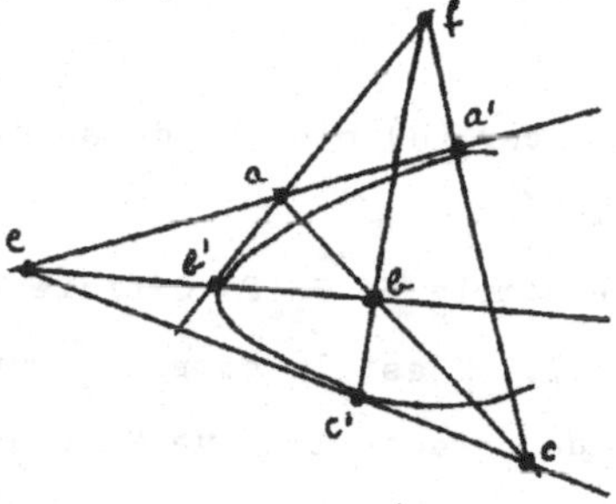

On a :

(1) a',b',c' alignés $\iff \begin{vmatrix} 1 & \lambda & 0 \\ 1 & 0 & \mu \\ 1 & \nu & \alpha\nu \end{vmatrix} = \mu\nu + \alpha\lambda\nu - \lambda\mu = 0$ (1)'

(2) ab',bc',ca' concourantes

$$\begin{vmatrix} \mu & 0 & -1 \\ -\nu & 1 & 0 \\ \alpha\lambda & -\alpha & 1 \end{vmatrix} = \mu - \alpha\nu + \alpha\lambda = 0 \qquad (2)'$$

De (1)' et (2)' résulte $\nu - \lambda = \frac{\lambda\nu}{\lambda - \nu}$ $\quad (\lambda \neq \nu$ car $\alpha(\nu - \lambda) = \mu \neq 0)$

soit : (R) $\lambda^2 - \lambda\nu + \nu^2 = 0$

Inversement supposons donné $\lambda, \nu \in K - \{0\}$ vérifiant la relation (R) Soit $\alpha \in K - \{0\}$ quelconque et posons $\mu = \lambda(\nu - \lambda)$. Alors $\mu \neq 0$ (car $\lambda = \nu$ est impossible d'après (R)) et $\lambda, \mu, \nu, \alpha$ vérifient (1)' et (2)'. Soit $\{e,a,b\}$ une base de $P_2(K)$; Posons :

$a' = e + \lambda a, \quad b' = e + \mu b, \quad c' = e + \nu c, \quad c = a + \alpha b$

et soit f le point d'intersection dans $P_2(K)$ (qui existe car (2)' $\Longrightarrow$ (2)) des droites ab', bc'; ca'. On vérifie alors que les 8 points ainsi définis sont distincts et que le matroïde induit par

$P_2(K)$ sur cet ensemble de points vérifie les hypothèses du lemme. D'après l'unicité c'est donc une représentation sur K de $\mathcal{M}(E)$.

Donc $\mathcal{M}(E)$ est représentable sur K si et seulement si (R) admet dans K une solution non nulle ou encore, l'équation étant homogène on peut poser $\lambda = 1$, si et seulement si l'équation algèbrique

(R') $\quad \nu^2 - \nu + 1 = 0$

admet une racine.

- $K = \mathbb{Q}$ (ou $\mathbb{R}$) est impossible car (R) n'a pas de solution réelle (par contre $K = \mathbb{C}$ est possible)[1].

- K = l'extension algèbrique simple de CG(2) définie par le polynome irréductible $\nu^2 - \nu + 1$ (i.e. K est le corps de rupture de $\nu^2 - \nu + 1$). Ce corps, fini, de caractéristique 2, convient par construction.

Notons aussi :

- K = CG(3) est possible : $\nu = 2$ vérifie (R') ce qui donne la représentation suivante :

$$e = (1,0,0), \quad a = (0,1,0), \quad b = (0,0,1), \quad c = (0,1,1)$$
$$a' = (1,1,0), \quad b' = (1,0,1), \quad c' = (1,2,2), \quad f = (1,2,1)$$

Plus généralement parmi les corps de caractéristique non nulle et impaire p les seuls qui conviennent sont ceux tels que :

$$p \equiv 1 \pmod 4 \quad \text{ou} \quad p \equiv 7 \pmod{12}$$

(résultats obtenus à partir de (R) en utilisant la théorie des résidus quadratiques).

---

(1) Par un papier d'Ingleton j'ai appris après coup que ce matroïde avait déjà été donné par Mac Lane comme exemple de matroïde représentable sur $\mathbb{C}$ mais pas sur $\mathbb{R}$.

Démonstration de la propriété II

ler Cas. Il existe un point e par lequel ne passe aucune ou une seule droite. Comme aux deux premiers cas du théorème I, on ajoute un point e' à une représentation de $\mathcal{M}$ (E-e) sur K = $\mathbb{Q}$ ou CG(2) ; lorsque K = CG(2), il est parfois nécessaire pour que e' existe de prendre une extension de CG(2) pour représenter $\mathcal{M}$(E-e). (afin qu'il reste assez de points pour trouver un e' satisfaisant aux conditions)

2ème Cas. Par tout point passe au moins 3 droites, et bien entendu on peut supposer le matroïde non binaire. Ce cas est résolu par le lemme 2.

3ème Cas. Par tout point passe au moins 2 droites et il existe un point e par lequel passe exactement 2 droites :

$$d_1 = \{e, a_1, b_1 \ldots\} \quad \text{et} \quad d_2 = \{e, a_2, b_2 \ldots\}$$

Comme les droites $d_1 - e$, $d_2 - e$, ne se coupent pas dans le sous-matroide $\mathcal{M}$(E-e), celui-ci d'ordre 7, n'est pas le matroïde de Fano et est donc représentable sur $\mathbb{Q}$ (en effet d'après la démonstration de I tout matroïde non binaire est représentable sur $\mathbb{Q}$. Par ailleurs tout matroïde binaire ne contenant pas comme mineur celui de Fano ou son dual est représentable sur tout corps). Soit donc une représentation de $\mathcal{M}$(E-e) dans $P_2(\mathbb{Q})$ (les points de celle-ci seront marqués avec un accent : ainsi a_1' pour $a_1 \in E$) et e' l'intersection dans $P_2(\mathbb{Q})$ des droites d_1' et d_2' (définies par a_1' b_1' et a_2' b_2'). Montrons que l'on peut toujours s'arranger pour n'avoir d'autre droite de la représentation de $\mathcal{M}$(E-e) passant par e', si bien qu'en lui adjoignant e' on obtient une représentation de $\mathcal{M}$(E).

Supposons que 2 points x',y' (différents de $a_1'b_1'$, et a_2',b_2') soient alignés avec e' sans l'être dans $\mathcal{M}$(E) puisque par e ne passe que 2 droites. Alors x' y' $\notin$ d'_1, d'_2 (en effet si par exemple x' $\in$ d'_1 alors e' a'_1 b'_1 x' y' seraient alignés aussi, e x y le seraient dans $\mathcal{M}$(E)) et la droite d' = {e', x', y'....} est unique en son genre (car |E'| < 9) si bien qu'il nous suffit de montrer qu'on peut éviter d'avoir e'x'y' alignés pour obtenir avec e' une représentation de $\mathcal{M}$(E). Enfin par x passe 2 droites (non triviales toujours) δ_1 et δ_2, par y passe λ_1 et λ_2; soit z le 8ème point du matroïde, z' son image dans la représentation.

1) z' $\in$ d'. Si 2 des 3 points x',y',z', ont 2 droites autres que d' par exemple x' et y', on a alors une représentation dans $P_2(\mathbb{Q})$ d'un matroïde d'ordre 7 (sur les points e' a'_1 b'_1 a'_2 b'_2 x' y') ayant 7 droites, ce qui est exclu par le lemme 1 (car c'est alors le matroïde de Fano). Donc x' et y', toujours par exemple, n'ont qu'une seule droite chacun autre que d', z' pouvant en avoir 2. La configuration est alors nécessairement celle de la figure 1 : une rotation de la droite x'y' autour de z' dans $P_2(\mathbb{Q})$ permet d'obtenir x" et y" non alignés avec e' sans rien changer aux autres droites.

2) z' $\notin$ d', d'_1, d'_2.

α) $|\delta_1| = |\delta_2| = |\lambda_1| = |\lambda_2| = 3$. L' un des 5 points a'_1 b'_1 a'_2 b'_2 z' n'a que 2 droites car sinon comme e' x' y' ont (avec d') 3 droites on aurait une représentation du matroïde du lemme 2 sur $\mathbb{Q}$.

- Supposons que ce soit par a'_1 (ou b'_1, a'_2, b'_2; idem) qui ait donc d'_1 et δ'. Si $\delta' \not\ni$ x',y', alors les 4 droites δ'_1, δ'_2, λ'_1 , λ'_2 seraient uniquement définies par les 4 points z', b'_1, a'_2, b'_2 ; comme par ailleurs $d'_2 \not\ni$ e' et la 6ème droite est z'$b'_1 \not\ni a'_1$ (car z' $\notin$ d'_1), on a la configuration de la fig. 2 et l'impossibilité de construire en a'_1 la droite δ'.

- Supposons que ce soit z' et que les 2 droites de z' ne contiennent ni x' ni y'. Sur E - z' on aurait alors une représentation de Fano dans $P_2(Q)$ ce qui est exclu.

Donc un des 4 points a'_1 b'_1 a'_2 b'_2 n'a que 2 droites et l'une d'elles passe par x' (ou y'). On peut alors déplacer x' de façon à n'avoir plus e' x' y' alignés.

- voir fig. 3 (a'_1 est le point en question).

β) On a par exemple $\delta_1 \geq 4$ et alors $|\delta_1| = 4$

Car une telle droite ne peut avoir plus que z',x' par exemple, un point de d'_1, un point de d'_2 ; soit $\delta_1 = \{z', a'_1, x', a'_2\}$. Une droite de y' doit contenir z' (sinon on retrouve Fano) par exemple z' b'_1 . Enfin l'autre ne peut être que a'_1 b'_2 . On a la configuration de la fig. 4 et là on peut déplacer y'.

3) $z' \in d'_1$ (ou d'_2 , idem). On raisonne comme en 2) α) (les δ_i et λ_i n'ont que 3 points) : a'_1 par exemple n'a que 2 droites ; l'une d'elles contient x' ou y' sinon on aurait Fano sur E- a'_1 etc ...

[Remarque : L'essentiel chaque fois est de trouver parmi l'un des 5 points a'_1 b'_1 a'_2 b'_2 z' un n'ayant que 2 droites dont l'une passe par x' ou y' et n'a que 3 points .]

Figures

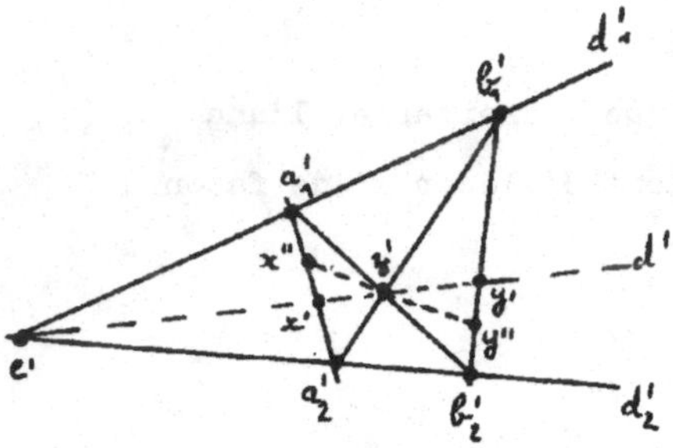

figure 1 (cas 1))

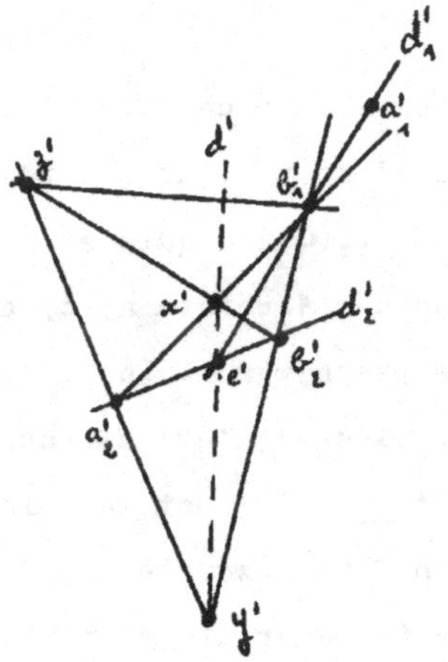

figure 2 (cas 2) α))

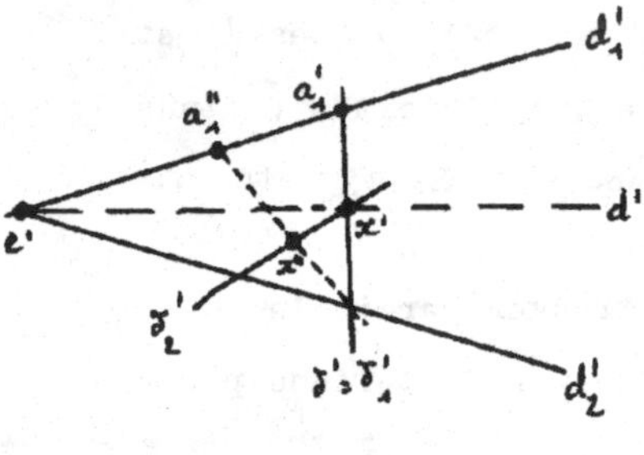

figure 3 (cas 2) α))

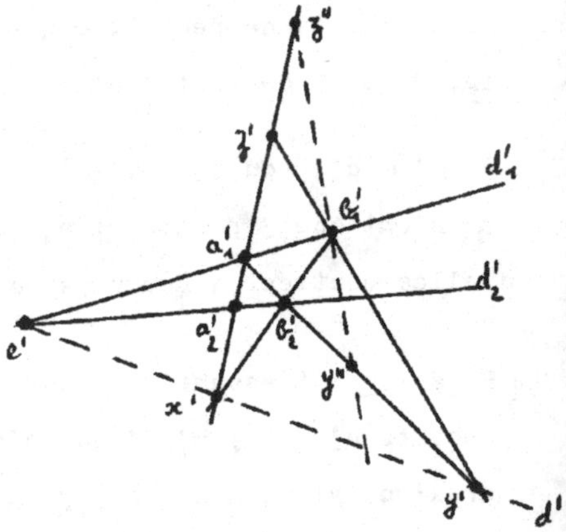

figure 4 (cas 2) β))

<u>Conclusion</u>. Tous les matroïdes géométriques d'ordre 7 sont représentables sur $\mathbb{Q}$ sauf celui de Fano et son dual qui le sont sur CG(2) Tous les matroides d'ordre 8 de rang $\neq$ 4 sont représentables sur un corps fini de caractéristique 2 ou sur $\mathbb{Q}$.

Bibliographie

[1] Lazarson, " The Représentation Problem for Independance Functions " .
J.Lond. Math. Soc. , Vol 33 (1958), p 21.

[2] Ingleton A.W. , " A Note of Independance Functions and rank ".
J.Lond. Math. Soc. , Vol 34 (1959), p 49.

[3] Fournier J.C. , " Sur la représentation sur un corps des matroïdes à 7 ét 8 éléments " .
C.R. Acad. Sc. Paris (ler Avril 70 série A).

CONDITIONS FOR REPRESENTABILITY AND TRANSVERSALITY OF MATROIDS

A.W. Ingleton

1. Of the many equivalent ways to describe a matroid I wish to concentrate, initially at least, on the aspect of rank.

Let S be a finite set. A rank function rk from the power set 2^S to the non-negative integers is characterized by the properties :

(R0) $\mathrm{rk}X \leqslant |X|$,

(R1) $X \subseteq Y \Rightarrow \mathrm{rk}X \leqslant \mathrm{rk}Y$,

(R2) $\mathrm{rk}(X \cap Y) + \mathrm{rk}(X \cup Y) \leqslant \mathrm{rk}X + \mathrm{rk}Y$,

for all X, $Y \subseteq S$. ($|X|$ denotes the cardinal of X.)

The pair $\mathfrak{M} = (S, \mathrm{rk})$ is called a matroid.

The following single inequality is equivalent to (R1) and (R2).

(R3) $\mathrm{rk}X_1 + \mathrm{rk}(X_1 \cup X_2 \cup X_3) \leqslant \mathrm{rk}(X_1 \cup X_2) + \mathrm{rk}(X_1 \cup X_3)$

for all X_1, X_2, $X_3 \subseteq S$.

The properties are, of course, suggested by vector space theory. In fact, however, the natural rank function for subsets of a vector space (over an arbitrary division ring) satisfies a more complicated inequality, with a recognizable family resemblance to (R3) but involving four subsets, namely

(R4) $\mathrm{rk}X_1 + \mathrm{rk}X_2 + \mathrm{rk}(X_1 \cup X_2 \cup X_3) + \mathrm{rk}(X_1 \cup X_2 \cup X_4) + \mathrm{rk}(X_3 \cup X_4)$
$\leqslant \mathrm{rk}(X_1 \cup X_2) + \mathrm{rk}(X_1 \cup X_3) + \mathrm{rk}(X_1 \cup X_4) + \mathrm{rk}(X_2 \cup X_3) + \mathrm{rk}(X_2 \cup X_4)$.

(For a proof see [2].)

It is easy to verify (by putting $X_4 = X_2$) that (R4) implies (R3), and hence (R1) and (R2). A matroid constructed by Vamos shows, however, that (R4) is not a consequence of (R0) - (R2). Indeed, it was his example which led me to look for (R4).

In the Vamos example, $S = X_1 \cup X_2 \cup X_3 \cup X_4$, where the X_i are disjoint sets of cardinal 2,

$$rk(X_i \cup X_j) = 3 \qquad (i = 1, 2 \;;\; i < j \leqslant 4).$$

and, for all other subsets of S, $rkX = \min(4, |X|)$.

A matroid is (linearly) representable if there exists a rank-preserving mapping of the underlying set S into a vector space. We may therefore state

Theorem 1 : *A necessary condition for a matroid to be representable is that the rank function satisfy* (R4).

Using the fact that the dual of a representable matroid is representable and hence satisfies (R4) one can obtain another necessary conditon for representability, namely

$$\text{(R4')} \quad \delta X_1 + \delta X_2 + \delta(X_1 \cap X_2 \cap X_3) + \delta(X_1 \cap X_2 \cap X_4) + \delta(X_3 \cap X_4)$$
$$\geqslant \delta(X_1 \cap X_2) + \delta(X_1 \cap X_3) + \delta(X_1 \cap X_4) + \delta(X_2 \cap X_3) + \delta(X_2 \cap X_4),$$

where $\delta X = |X| - rkX$ is the dependence of a subset X. I do not know whether (R4') is implied by (R0) and (R4).

It is interesting to speculate whether there are any independent higher order rank inequalities, involving more than four subsets, satisfied in representable matroids. It would not be surprising if (R3) and (R4) were the first two members of an infinite hierarchy of inequalities.

2. I now turn to transversal matroids ; first some notation :

$\mathcal{A}_* = (A_i : i \in I)$ denotes a finite family of subsets of S ;

$\mathcal{A}^* = (A^i : i \in I)$, where $A^i = S \setminus A_i$, denotes the complementary family to $\mathcal{A}_*$;

for any $J \subseteq I$, we write

$$A_J = \bigcup_{i \in J} A_i ,$$

$$A^J = \bigcap_{i \in J} A^i = S \setminus A_J \ (= S \text{ if } J = \emptyset),$$

$$J' = I \setminus J.$$

A subset X of S is a <u>partial transversal</u> (PT) of $\mathcal{A}_*$ if there is an injective mapping $f : X \to I$ such that, for each $x \in X$, $x \in A_{fx}$.

By P. Hall's theorem, X is a PT if and only if

(T1) $|X \cap A_J| \geq |J| + |X| - |I|$ for all $J \subseteq I$.

The key lemma for the sequel is a complementary form of Hall's condition which follows by a simple calculation.

<u>Lemma 1</u> : *X is a PT of $\mathcal{A}_*$ if and only if*

(T1') $|X \cap A^{J'}| \leq |J|$ *for all* $J \subseteq I$.

The pleasant feature of (T1'), apart from its convenience for my purpose, is that it is still a criterion for a PT in the case when S and I are infinite, provided that $\mathcal{A}_*$ is such that each element of S belongs to only finitely many A_i. Although I shall confine attention to the finite case here, my characterization of transversal matroids does in fact generalize to the infinite case.

It is well known that the PTs of any family $\mathcal{A}_*$ are the independent sets of a matroid ; by (T1) the associated rank function is given by

$$\text{rk}X = \min_{J \subseteq I} (|X \cap A_J| + |J'|).$$

The resulting matroid will be denoted by $\mathcal{M}(\mathcal{A}_*)$. Any matroid $\mathcal{M}$ which is so obtainable is called a <u>transversal matroid</u> and any $\mathcal{A}_*$ for which $\mathcal{M}(\mathcal{A}_*) = \mathcal{M}$ is called a <u>presentation</u> of $\mathcal{M}$. A fairly easy consequence of Lemma 1 is :

Lemma 2 : *Let* $\mathfrak{M} = (S, rk)$ *be a matroid and* $\mathcal{A}_* = (A_i : i \in I)$ *be a finite family of subsets of* S. *Then* $\mathfrak{M} = \mathfrak{M}(\mathcal{A}_*)$ *if and only if the following two properties are satisfied.*

(i) $rkA^{J'} \leq |J|$ *for all* $J \subseteq I$.

(ii) *For every circuit* C *of* $\mathfrak{M}$ *there exists* $J \subseteq I$ *such that*

$$C \subseteq A^{J'}, \quad |J| = rkC.$$

[A circuit is a minimal dependent set, that is, $rkC = |C| - 1 = rk(C \setminus \{x\})$ for all $x \in C$.]

Definition : In a matroid (S, rk) with $rkS = r$, a hyperplane H is a subspace of rank $r - 1$ (or $(r - 1)$-flat), that is,

$$rkH = r - 1, \quad rk(H \cup \{x\}) = r \quad \text{for all} \quad x \notin H.$$

A quasisimplex is a family of hyperplanes

$$\mathcal{H}^* = (H^i : i \in I), \quad |I| = r,$$

such that

$$rkH^{J'} \leq |J| \quad \text{for all} \quad J \subseteq I.$$

For each $J \subseteq I$ we call $H^{J'}$ a k-face of $\mathcal{H}^*$, where $k = |J|$.

Theorem 2 : *A matroid is transversal if and only if it contains a quasisimplex* $\mathcal{H}^*$ *such that each circuit* C *is contained in some* k*-face of* $\mathcal{H}^*$ *with* $k = rkC$.

Theorem 2 follows immediately from Lemma S and a theorem in a forthcoming paper of Bondy and Welsh [1] that every transversal matroid has a presentation by a family of r cocircuits, $r = rkS$. [A cocircuit is the complement of a hyperplane.] In fact, I have an alternative direct proof of Theorem 2 from Lemma 2 valid also in the infinite case [3].

Several known properties of transversal matroids seem to follow relatively easily from Theorem 2, in particular the corollary.

<u>Corollary</u> (Piff and Welsh [4]): *A transversal matroid is representable over every infinite (or sufficiently large finite) division ring.*

Each $x \in S$ belongs to a unique smallest face $H^{J'}$ of $\mathcal{H}^*$, we assign a coordinate vector $\underset{\sim}{\xi}(x) = (\xi_1, \ldots, \xi_r)$ to x with $\xi_i \neq 0$ if and only if $i \in J$. It is then only necessary to ensure that there are no "accidental" relations of dependence between the vectors $\underset{\sim}{\xi}(x)$ $(x \in S)$.

REFERENCES

[1] J.A. Bondy and D.J.A. Welsh, "Some theorems on the structure of matroids" to be published.

[2] A.W. Ingleton, "Representation of matroids", Proc. Of the Conference on Combinatorial Mathematics and its Applications, Oxford (1969).

[3] A.W. Ingleton, "A geometrical characterization of transversal independence structures", Bull. Lond. Math. Soc.

[4] M.J. Piff and D.J. Welsh, "On the vector representation of matroids", J. London Math. Soc. (2) 2 (1970) 284-8.

SUR LA DUALITE EN THEORIE DES MATROIDES

Michel Las Vergnas

Soit M un matroïde sur un ensemble E. Lorsque E est fini on définit à partir de M le matroïde dual M* de M [1] , [4] : M satisfait alors à la relation M = M**. Dans le cas général (E fini ou infini) on peut définir un matroïde M* se réduisant au matroïde dual dans le cas fini, mais on n'a plus nécessairement M = M**.

Dans ce qui suit on étudie des conditions nécessaires et suffisantes pour que cette propriété soit vérifiée [2].

Soit E un ensemble (fini ou infini) et soit M un <u>matroïde</u> sur E (" φ-espace" [3] , "pregeometry" [1]). Nous désignerons par

$\mathcal{L} \subset \mathcal{P}(E)$ (1) l'ensemble des <u>parties libres</u> de M ("independent sets"[1])

$\mathcal{B} \subset \mathcal{P}(E)$ l'ensemble des <u>bases</u> de M

$\mathcal{S} \subset \mathcal{P}(E)$ l'ensemble des <u>stigmes</u> de M ("atom" [4], "circuit" [1] , [3]).

$r\ \mathcal{P}_f(E) \to \mathbb{N}$ (1) la <u>fonction rang</u> de M (au sens de [1], [3] ; soit t le rang au sens de [4] : pour tout X fini $\subset$ E on a $r(X) + t(M \times X) = |X|$).

$\varphi\ \mathcal{P}(E) \to \mathcal{P}(E)$ la <u>fonction génératrice</u> de M (cf. [3] ; dans la terminologie de [1] $\varphi(X) = \overline{X}$ est le "span" de $X \subset E$)

$\mathcal{F} \subset \mathcal{P}(E)$ l'ensemble des <u>fermés</u> de M ("closed sets" [1])

$\mathcal{H} \subset \mathcal{P}(E)$ l'ensemble des <u>hyperplans</u> de M ("copoints" [1]).

---

(1) $X \subset Y$ est une abréviation de $\forall x\ (x \in X \Longrightarrow x \in Y)$
$\mathcal{P}_f(E)$ est l'ensemble des parties finies de E , $\mathbb{N}$ l'ensemble des entiers naturels.

La donnée de l'un quelconque de $\mathcal{L}, \mathcal{B}, \mathcal{S}$, r, φ, $\mathcal{F}$ ou $\mathcal{H}$ satisfaisant à une axiomatique convenable (cf. [1] , [3] , [4]) définit un matroïde sur E.

<u>Exemple [4]</u> : (SO), (S1), (S2) constituent une axiomatique de l'ensemble des stigmes d'un matroïde

(SO) $\emptyset \notin \mathcal{S}$ et $(X \in \mathcal{S} \Rightarrow X$ est fini$)$.

(S1) $X,Y \in \mathcal{S} \Rightarrow X \not\subset Y$.

(S2) pour tous $X,Y \in \mathcal{S}$ et $x \in X \cap Y$, $y \in Y-X$ il existe $Z \in \mathcal{S}$ tel que $y \in Z \subset (X \cup Y) - x$.

$\mathcal{L}, \mathcal{B}, \mathcal{S}, r, \varphi, \mathcal{F}, \mathcal{H}$ sont liés par les relations suivantes :

(BL) $\mathcal{L} = \{X | X \subset E , \exists B \in \mathcal{B} \ X \subset B\}$

(SL) $\mathcal{L} = \{X | X \subset E , \forall Y \subset X \ Y \notin \mathcal{S}\}$

(RL) $\mathcal{L} = \{X | X \subset E , \forall Y$ fini $\subset X \ r(Y) = |Y|\}$

(GL) $\mathcal{L} = \{X | X \subset E , \forall x \in X \ x \notin \varphi(X-x)$

(LB) $\mathcal{B} = \{X | X \in \mathcal{L}$, maximal pour l'inclusion avec cette propriété$\}$

(LS) $\mathcal{S} = \{X | X \in \mathcal{P}(E) - \mathcal{L}$, minimal pour l'inclusion avec cette propriété$\}$

(LR) pour X fini $\subset E$, $r(X) = \sup |Y| \ \{Y \subset X , Y \in \mathcal{L}\}$

(BR) pour X fini $\subset E$, $r(X) = \sup\limits_{B \in \mathcal{B}} |X \cap B|$

(LG) pour $X \subset E \ \varphi(X) = X \cup \{x | x \in E \ \exists Y \subset X \ \ Y \in \mathcal{L}, Y+x \notin \mathcal{L}\}$.

(SG) pour $X \subset E \ \varphi(X) = X \cup \{x | x \in E \ \exists \ Y \subset X \ \ Y+x \in \mathcal{S}\}$

(GF) $\mathcal{F} = \{X | X \subset E , \ \varphi(X) = X\}$

(FG) pour $X \subset E , \varphi(X) = \bigcap\limits_{Y \supset X , Y \in \mathcal{F}} Y$

(FH) $\mathcal{H} = \{X | X \in \mathcal{F}, X \neq E$, maximal pour l'inclusion avec ces propriétés$\}$

(HF) $\mathcal{F} = \{ \bigcap\limits_{X \in \mathfrak{X}} X | \mathfrak{X} \subset \mathcal{H}\}$.

Rappelons quelques propriétés classiques des matroïdes :

1) Soit $X \subset E$; (i), (ii), (iii), sont équivalents :

(i) Y est maximal pour l'inclusion avec les 2 propriétés $Y \subset X$ et $Y \in \mathcal{L}$.

(ii) Y est minimal pour l'inclusion avec les 2 propriétés $Y \subset X$ et $\varphi(Y) \supset X$.

(iii) $Y \subset X$, $Y \in \mathcal{L}$ et $\varphi(Y) \supset X$.

Nous dirons que $Y \subset X$ avec une et donc les 3 des propriétés (i), (ii), (iii) est <u>une base de</u> X. Théorème de la base incomplète : pour tous Y', Y'' tels que $Y' \subset Y'' \subset X$, $Y' \in \mathcal{L}$ et $\varphi(Y'') \supset X$ il existe une base Y de X telle que $Y' \subset Y \subset Y''$. D'autre part, toutes les bases de X ont même cardinal : en particulier si X est fini $|Y| = r(X)$ pour toute base Y de X.

2) Soient X,Y finis $\subset E$: $r(X \cup Y) + r(X \cap Y) \leq r(X) + r(Y)$ (sous-modularité du rang) et pour $X \subset Y$ $r(X) \leq r(Y)$.

3) Soit $X \subset E$: $\varphi(X) = \bigcup_{Y \text{ fini} \subset X} \varphi(Y)$, $\varphi(\varphi(X)) = \varphi(X) \supset X$ et si $x \notin \varphi(X)$, $x \in \varphi(X+y)$ on a $y \in \varphi(X+x)$.

4) Si $X \in \mathcal{F}$ et $x \notin X$ il existe $H \in \mathcal{H}$ tel que $H \supset X$ et $x \notin H$.

§ 1.- LE MATROIDE M*.

<u>Lemme 1.1</u>. <u>Soit</u> $X \subset E$: <u>on a</u> $E\text{-}X \in \mathcal{F} \Longleftrightarrow |X \cap S| \neq 1 \ \forall S \in \mathcal{S}$.

<u>Démonstration</u> : $E\text{-}X \notin \mathcal{F} \Longleftrightarrow$ il existe $x \in X \cap \varphi(E\text{-}X)$ (GF)

$\Longleftrightarrow$ il existe $x \in X$ et $S \in \mathcal{S}$ tels que $x \in S \subset (E\text{-}X)+x$ (SG)

$\Longleftrightarrow$ il existe $S \in \mathcal{S}$ $|S \cap X| = 1$.

<u>Définition 1</u> : Posons $\mathcal{S}^* = \{X | X$ <u>fini</u> $\subset E$ <u>tel que</u> $E\text{-}X \in \mathcal{H}\}$ équivalent par le lemme à $\mathcal{S}^* = \{X | X$ fini $\subset E$, $X \neq \emptyset$, tel que $|X \cap S| \neq 1$ $\forall S \in \mathcal{S}$ et minimal pour l'inclusion avec cette propriété$\}$(1).

<u>Proposition 1.2</u>. $\mathcal{S}^*$ <u>vérifie les axiomes (S0), (S1), (S2) et définit donc une structure de matroïde</u> $M^*(\mathcal{L}^*, \mathcal{B}^*, \mathcal{S}^*, r^*, \varphi^*, \mathcal{F}^*, \mathcal{H}^*)$ <u>sur</u> E.

La démonstration de la proposition 1.2. donnée par [4] dans le cas fini (2.61) à partir de la deuxième définitinn de $\mathcal{S}^*$ s'étend sans changement au cas général. Lorsque E est fini M^* est le <u>matroïde dual</u> de M.

<u>Proposition 1.3</u>.

1). $\mathcal{L}^* = \{X | X \subset E$, $\forall Y$ fini $\subset X$ $\varphi(E\text{-}Y) = E\}$

$\mathcal{L}^* = \{X | X \subset E$, $\forall Y$ fini $\subset X$ $\exists B \in \mathcal{B}$ $B \cap Y = \emptyset\}$

---

(1) On a également $\mathcal{S}^* = \{X | X$ fini $\subset E, X \neq \emptyset$ tel que 1) $|X \cap S| \neq 1$ $\forall S \in \mathcal{S}$, 2) si $|X| \geq 2$, $\forall x,y \in X$, $x \neq y$ $\exists S \in \mathcal{S}$ $X \cap S = \{x,y\}\}$.

2). pour X fini $\subset E$ $r^*(X) = \lim_{Y \text{ fini} \subset X} (|X| + r(Y-X) - r(Y))$

3). pour $X \subset E$ $\varphi^*(X) = X \cup \{x | x \in E$ tel qu'il existe Y fini $\subset X$ tel que $x \notin \varphi((E-Y)-x)$

Démonstration : 1) L'équivalence des 2 expressions de $\mathcal{L}^*$ est immédiate par le théorème de la base incomplète. Posons $\mathcal{L}' = \{X | X \subset E, \forall Y$ fini $\subset X\ \varphi(E-Y) = E\}$. Par (SL), $\mathcal{L}^* = \{X | X \subset E, \forall Y \subset X\ Y \notin \mathcal{J}^*\}$. $\mathcal{L}' \subset \mathcal{L}^*$: soit $X \in \mathcal{L}'$ et supposons $X \notin \mathcal{L}^*$ i.e. il existe $Y \subset X$, $Y \in \mathcal{J}^*$. Y est fini d'où $\varphi(E-Y) = E$. $Y \neq \emptyset$; soit $x \in Y$: il existe $Z \subset E-Y$ tel que $Z+x \in \mathcal{J}$ (SG) d'où $(Z+x) \cap Y = \{x\}$ contredisant $Y \in \mathcal{J}^*$. Inversement $\mathcal{L}^* \subset \mathcal{L}'$: il suffit de montrer que X fini $\in \mathcal{L}^*$ implique $X \in \mathcal{L}'$. Supposons au contraire $\varphi(E-X) \neq E$ i.e. $Y = X - \varphi(E-X) \neq \emptyset$. Pour tout $S \in \mathcal{J}$ on a $|S \cap Y| \neq 1$ (en effet si $S \cap Y = \{x\}$ pour $S \in \mathcal{J}$ on a $S-x \subset E-X$ d'où $x \in \varphi(E-X)$ (SG)), par suite il existe $Y_o \subset Y \subset X$, $Y_o \in \mathcal{J}^*$ ce qui contredit $X \in \mathcal{L}^*$ (SL).

2). Soit X fini $\subset E$: $\lim_{Y \text{ fini} \subset X} (|X| + r(Y-X) - r(Y))$ existe :

en effet soit Y' fini, $Y' \supset Y \supset X$, on a par la sous-modularité de r $r(Y') + r(Y-X) = r((Y'-X) \cup Y) + r((Y'-X) \cap Y) \leq r(Y'-X) + r(Y)$ i.e. $|X| + r(Y-X) - r(Y) \leq |X|$ est une fonction croissante de Y.

Lemme 1.4. Soit X fini $\subset E$ et soit A_o une base de $E-X$, A_o se prolonge en une base B_o de E et on a :

(i) $|X - B_o| = \lim_{Y \text{ fini} \subset X} (|X| + r(Y-X) - r(Y)) = \sup_{B \in \mathcal{B}} |X - B|$

(ii) $X - B_o$ est une base de X relativement à M^*.

(i) Soit B une base de E relativement à M et soit $X \subset E$. Posons $\sigma(X) = \sigma_B(X) = \{x | x \in B, X \notin \varphi(B-x)\}$.

a) $\sigma(X) = \bigcup_{x \in X} \sigma(x)$. En particulier si X est fini $\sigma(X)$ est fini (pour $x \notin B$ il existe B_x fini $\subset B$ tel que $x \in \varphi(B_x)$ d'où $\sigma(x) \subset B_x$, pour $x \in B$ $\sigma(x) = \{x\}$).

$\sigma(X) \supset B \cap X$.

b) $\varphi(\sigma(X)) \supset X$ (pour $x \notin B$ $\sigma(x)+x \in \mathcal{J}$), $\sigma(X) = \bigcap_{Y \subset B,\ \varphi(Y) \supset X} Y$.

En particulier $\sigma(X)$ est une base de $X \cup \sigma(X) = \tilde{X}$.

On a $|X-B| = |X| + |\sigma(X) - X| - |\sigma(X)| = |X| + r(\tilde{X} - X) - r(\tilde{X})$
$\leq \lim_{Y \text{ fini} \subset X} (|X| + r(Y-X) - r(Y))$.

Inversement soit B_o une base de E telle que $A_o = B_o - X$ soit une base de E-X et soit Y_o fini $\subset X$ tel que

$$|X| + r(Y_o - X) - r(Y_o) = \lim_{Y \text{ fini} \subset X} (|X| + r(Y-X) - r(Y)).$$

Posons $\tilde{Y}_o = Y_o \cup \sigma(Y_o)$: $\tilde{Y}_o$ est fini et $\tilde{Y}_o \supset Y_o$ d'où

$$|X| + r(\tilde{Y}_o - X) - r(\tilde{Y}_o) = \lim_{Y \text{ fini} \subset X} (|X| + r(Y-X) - r(Y)).$$

D'autre part $\sigma(Y_o)$ est une base de $\tilde{Y}_o$ et $\sigma(Y_o) - X = \sigma(Y_o - X)$ est une base de $\tilde{Y}_o - X$: en effet A_o étant une base de E-X pour $x \in Y_o - X$ on a $\sigma(x) \subset A_o \subset E-X$ d'où $\sigma(Y_o - X) \subset E-X$. Par suite

$$|X| + r(\tilde{Y}_o - X) - r(\tilde{Y}_o) = |X| + |\sigma(Y_o) - X| - |\sigma(Y_o)| = |X - \sigma(Y_o)| = |X - B_o|.$$

(ii) $X - B_o \in \mathcal{L}^*$ par 1). Soit $x \in B_o \cap X$: posons $S' = \{y \mid y \in X$ $\sigma(y) \ni x\} = X - \varphi(B_o - x) \subset (X - B_o) + x$. $A_o \subset B-x$ étant une base de E-X , $E-X \subset \varphi(B-x)$ d'où $E - S' = \varphi(B-x) \in \mathcal{H}$. S' étant fini, $S' \in \mathcal{S}^*$ d'où $(X - B_o) + x \notin \mathcal{L}^*$ (SL) et donc $X - B_o$ est une base de X relativement à M*.

3) Soit $X \subset E$, posons $\varphi'(X) = X \cup \{x \mid x \in E$ t.q. il existe Y fini $\subset X$ tel que $x \notin \varphi((E-Y)-x)\}$. Soit $x \in \varphi^*(X) - X$: il existe $Y \subset X$ tel que $Y + x \in \mathcal{S}^*$ (SG). On a $x \notin \varphi((E-Y)-x)$ (sinon il existe $Z \subset (E-Y)-x$ tel que $Z + x \in \mathcal{S}$ d'où $(Z+x) \cap (Y+x) = \{x\}$), Y étant fini on a bien $x \in \varphi'(X)$.

Inversement soit $x \in \varphi'(X) - X$: il existe Y fini $\subset X$ tel que $x \notin \varphi((E-Y)-x)$. Supposons Y choisi avec cette propriété de façon que $|Y|$ soit minimal. $Y + x \notin \mathcal{L}^*$ en effet $x \notin \varphi(E-(Y+x))$ (cf. 1)). D'autre part, par la minimalité de $|Y|$, pour tout $y \in Y$ on a $x \in \varphi((E-(Y-y))-x) = \varphi((E-Y)-x)+y)$ d'où $y \in \varphi(E-Y)$ i.e. $\varphi(E-Y) = E$, $Y \in \mathcal{L}^*$ et par (LG) $x \in \varphi^*(X)$.

<u>Corollaire : Si φ et ψ sont les fonctions génératrices de 2 matroïdes sur</u> E <u>telles que</u> $\varphi(X) \subset \psi(X)$ <u>pour tout</u> $X \subset E$ <u>alors</u> $\varphi^*(X) \supset \psi^*(X)$ <u>pour</u> tout $X \subset E$.

Notons que par contre $r(X) \leq s(X)$ pour tout X fini $\subset E$ n'entraîne pas que r* et s* soient comparables.

§ 2. PAIRES MODULAIRES DANS UN MATROIDE M ($\mathcal{L}$,...)

Lemme 2.1. Soient $X,Y \subset E$, $B \subset X \cup Y$; (i) et (ii) sont équivalents :

(i) B est une base de $X \cup Y$, $B \cap X$ une base de X , $B \cap Y$ une base de Y et $B \cap X \cap Y$ une base de $X \cap Y$.

(ii) $B \in \mathcal{L}$, $B \cap X$ est une base de X et $B \cap Y$ une base de Y.

Démonstration : (ii) $\Longrightarrow$ (i) : il est clair que B est une base de $X \cup Y$. Montrons que $B \cap X \cap Y$ est une base de $X \cap Y$. En effet sinon il existe $x \in (X \cap Y)-B$ tel que $(B \cap X \cap Y) + x \in \mathcal{L}$. $B \cap X$ est une base de X : il existe $R \in \mathcal{S}$ tel que $x \in R \subset (B \cap X) + x$. $R-Y \neq \emptyset$ (car $(B \cap X \cap Y)+x \in \mathcal{L}$). De même il existe $S \in \mathcal{S}$ tel que $x \in S \subset (B \cap Y)+x$. $R \neq S$ d'où par (S2) il existe $T \in \mathcal{S}$ tel que $T \subset (R \cup X)-x \subset B$ contredisant $B \in \mathcal{L}$.

Définition 2. Nous dirons que X,Y constituent une paire modulaire (nous écrirons "(X,Y) est modulaire") s'il existe $B \subset X \cup Y$ avec l'une des propriétés équivalentes (i), (ii) du lemme 2.1. (cf.[1]).

Lemme 2.2. Lorsque X,Y sont finis (i) (X,Y) est modulaire et (ii) $r(X \cup Y)+r(X \cap Y) = r(X)+r(Y)$ sont équivalents.

Proposition 2.3. Soient $X,Y \subset E$; (i) et (ii) sont équivalents :

(i) (X,Y) est modulaire

(ii) tout $B \subset X \cup Y$ tel que B soit une base de $X \cup Y$ et $B \cap X \cap Y$ une base de $X \cap Y$ est tel que $B \cap X$ soit une base de X , $B \cap Y$ une base de Y.

Démonstration : (ii) $\Longrightarrow$ (i). Soit B_o' une base de $X \cap Y$, B_o' se prolonge en une base $B_o \supset B_o'$ de $X \cup Y$. On a $B_o \cap X \cap Y = B_o'$; par (ii) $B_o \cap X$ est une base de X , $B_o \cap Y$ une base de Y i.e. (X,Y) est modulaire.

(i) $\Longrightarrow$ (ii). Soit $B_o \subset X \cup Y$ tel que $B_o \in \mathcal{L}$, $B_o \cap X$ soit un base de X et $B_o \cap Y$ une base de Y, et soit $B \subset X \cup Y$ une base de $X \cup Y$ telle que $B \cap X \cap Y$ soit une base de $X \cap Y$.

Supposons par exemple que $B \cap X$ ne soit pas une base de X : il existe $x \in X-B$ tel que $(B \cap X)+x \in \mathcal{L}$. $B \cap X \cap Y$ étant une base de $X \cap Y$ $x \notin X \cap Y$ i.e. $x \in X-Y$. B est une base de $X \cup Y$: il existe un stigme $S \in \mathcal{S}$ tel que $x \in S \subset B+x$.

Soit alors $S \in \mathcal{S}$ tel que 1) $x \in S$, $S-Y \subset B+x$
2) $|S-X-B_o|$ soit minimal pour les $S \in \mathcal{S}$ ayant la propriété 1)
3) $|(S \cap X \cap Y)-B|$ soit minimal pour les $S \in \mathcal{S}$ ayant les propriétés 1) et 2). On a

1). $|S-X-B_o| = 0$. En effet soit au contraire $y \in S-X_o-B$. $B_o \cap Y$ est une base de Y : il existe $T \in \mathcal{S}$ tel que $y \in T \subset (B_o \cap Y)+y$. $x \notin T$: par (S2) il existe $S' \in \mathcal{S}$ tel que $x \in S' \subset (S \cup T)-y$. On a $S'-Y \subset S-Y \subset B+x$ et $|S'-X-B_o| < |S-X-B_o|$ ce qui contredit la définition de S.

2). $S-X = \emptyset$. Par 1) $S \subset X \cup B_o$; soit au contraire $y \in S-X \subset B_o$. On a $\varphi(B_o-y) \supset \varphi(B_o \cap X) \supset X$ d'où $y \in S \subset X \cup B_o \subset \varphi(B_o-y)+y$ i.e. $y \in \varphi(\varphi(B_o-y)) = \varphi(B_o-y)$ ce qui contredit $B_o \in \mathcal{L}$.

3). $|(S \cap X \cap Y)-B| = 0$. Soit au contraire $y \in (S \cap X \cap Y)-B$. $B \cap X \cap Y$ est une base de $X \cap Y$: il existe $T \in \mathcal{S}$ tel que $y \in T \subset (B \cap X \cap Y)+y$. $x \notin T$: par (S2) il existe $S' \in \mathcal{S}$ tel que $x \in S' \subset (S \cup T)-y$. Par 2) $S \subset X$ d'où $S'-Y \subset S-Y \subset B+x$ et $|(S' \cap X \cap Y)-B| < |(S \cap X \cap Y)-B|$ ce qui contredit la définition de S.

On a ainsi $S \subset (B \cap X)+x$ ce qui contredit $(B \cap X)+x \in \mathcal{L}$.

<u>Corollaire 1:</u> <u>Soit</u> (X,Y) <u>une paire modulaire</u> et $Z \supset X \cap Y$ <u>alors</u> $(X \cap Z, Y \cap Z)$ <u>est une paire modulaire</u>.

<u>Démonstration</u> : Soit B' une base de $X \cap Y$, B' se prolonge en une base C de $Z \cap (X \cup Y)$ qui se prolonge à son tour en une base B de $X \cup Y$. On a $B \cap Z = C$, $C \cap (X \cap Y) = B'$. Par la proposition 2.3 $B \cap X$ est une base de X , $B \cap Y$ une base de Y. Supposons que $C \cap X$ par exemple ne soit pas une base de $Z \cap X$: il existe $x \in (Z \cap X)-C$ tel que $(C \cap X)+x \in \mathcal{L}$. $B \cap X$ étant une base de X il existe $R \in \mathcal{S}$ tel que $x \in R \subset (B \cap X)+x$. C étant une base de $Z \cap (X \cup Y)$ il existe $S \in \mathcal{S}$ tel que $x \in S \subset C+x$. Comme $(B \cap Z \cap X)+x = (C \cap X)+x \in \mathcal{L}$ on a $R \not\subset Z$ d'où $R \neq S$: par (S2) il existe $T \in \mathcal{S}$ $T \subset (R \cup S)-x \subset B$ une contradiction.

<u>Corollaire 2</u> : <u>Soit</u> (X,Y) <u>une paire modulaire telle que</u> $X \cap Y = \emptyset$ <u>alors pour tout</u> $X' \subset X$, $Y' \subset Y$ (X',Y') <u>est modulaire</u>.

Corollaire 3 : Soient $X, Y \subset E$ tels que $X \cap Y = \emptyset$; (i), (ii), (iii), (iv), (v), (vi) sont équivalents :

(i) (X,Y) est modulaire

(ii) pour tous X', Y' finis, $X' \subset X$, $Y' \subset Y$ (X',Y') est modulaire

(iii) pour tous X', Y', $X' \subset X$, $Y' \subset Y$, tels que $X', Y' \in \mathcal{L}$ on a $X' \cup Y' \in \mathcal{L}$.

(iv) si B est une base de X, C une base de Y alors $B \cup C$ est une base de $X \cup Y$.

(v) quel que soit $S \in \mathcal{S}$ tel que $S \cap X \neq \emptyset$ et $S \cap Y \neq \emptyset$ on a $S-(X \cup Y) \neq \emptyset$.

(vi) pour tous X', Y' finis, $X' \subset X$, $Y' \subset Y$, on a $r(X' \cup Y') = r(X') + r(X') + r(Y')$.

§ 3. COMPARAISON DE M ET M**

Lemme 3.1. 1) la fonction rang r^{**} de M^{**} est donnée par

$$r^{**}(X) = \lim_{Y \text{ fini} \supset X} \left(\lim_{Z \text{ fini}, Z \cap Y = \emptyset} (r(X \cup Z) - r(Z)) \right)$$

$$= \sup_{Y \text{ fini} \supset X} \left(\sup_{Z \text{ fini}, Z \cap Y = \emptyset} (r(X \cup Z) - r(Z)) \right)$$

pour X fini $\subset E$.

2). la fonction génératrice φ^{**} de M^{**} est donnée par

$$\varphi^{**}(X) = \bigcup_{Y \text{ fini} \subset X} \left(\bigcap_{Z \supset Y, E-Z \text{ fini}} \varphi(Z) \right) \quad \text{pour } X \subset E.$$

Démonstration :

1). Soit X fini $\subset E$. Par la proposition 1.3.

$$r^{**}(X) = \lim_{Y \text{ fini} \subset X} (|X| + r^*(Y-X) - r^*(Y))$$

$$= \sup_{Y \text{ fini} \subset X} (|X| + r^*(Y-X) - r^*(Y))$$

$$r^*(Y-X) - r^*(Y) = \lim_{Z' \text{ fini} \subset Z-X} (Y-X + r(Z'-(Y-X)) - r(Z'))$$

$$- \lim_{Z' \text{ fini} \subset Y} (|Y| + r(Z'-Y) - r(Z'))$$

$$= -|X| + \lim_{Z' \text{ fini} \subset Y} (r(X \cup (Z'-Y)) - r(Z'-Y)).$$

La première expression s'obtient alors en posant $Z = Z'-Y$, la seconde en remarquant que $r(X \cup Z)-r(Z)$ est une fonction décroissante de Z.

2). Soit $X \subset E$. $\varphi^{**}(X) = \bigcup_{Y \text{ fini} \subset X} \varphi^{**}(Y)$

$$\varphi^{**}(Y) = Y \cup \{x \; x \in E \quad x \notin \varphi^*((E-Y)-x)\}$$

$x \notin \varphi^*((E-Y)-x) \Longleftrightarrow \forall Z' \text{ fini} \subset (E-Y)-x \quad x \in \varphi((E-Z')-x)$.

Posons $Z = (E-Z')-x$

$x \notin \varphi^*((E-Y) \Longleftrightarrow \forall Z \supset Y$ tel que $E-Z$ fini $x \in \varphi(Z)$ d'où

$$\varphi^{**}(X) = X \cup \bigcup_{Y \text{ fini} \subset X} \Big(\bigcap_{Z \supset Y, E-Z \text{ fini}} \varphi(Z)\Big).$$

Lorsque E est fini le lemme 3.1. donne $r^{**}(X) = r(X)$ pour tout X fini $\subset E$ i.e. $M = M^{**}$ (cf. prop. 2. 65 de [4]). Dans le cas général on a :

<u>Lemme 3.2.</u>

1). <u>Pour</u> X <u>fini</u> $\subset E$ $r^{**}(X) \leq r(X)$ (d'où $\mathcal{L}^{**} \subset \mathcal{L}$).

2). <u>Pour</u> $X \subset E$ $\varphi^{**}(X) \supset \varphi(X)$ (d'où $\mathcal{F}^{**} \subset \mathcal{F}$).

<u>Proposition 3.3.</u> <u>Pour tout matroïde M, $M^* = M^{***}$.</u>

<u>Démonstration</u> : Par le lemme 3.2. $\varphi^* \subset (\varphi^*)^{**} = \varphi^{***}$ et $\varphi \subset \varphi^{**}$ d'où par le corollaire de la proposition 1.3 $\varphi^* \supset (\varphi^{**})^* = \varphi^{***}$ i.e. $\varphi^* = \varphi^{***}$.

<u>Proposition 3.4.</u> <u>Soit M un matroïde sur un ensemble E ; (i), (ii), (iii) sont équivalents</u> :

<u>(i) $M = M^{**}$.</u>

<u>(ii) quel que soit</u> X <u>fini</u> $\subset E$ <u>il existe</u> Y <u>fini</u> $\supset X$ <u>tel que</u> $(X,E-Y)$ <u>soit modulaire.</u>

(iii) <u>quel que soient</u> X <u>fini</u> $\subset E$ <u>et</u> $x \notin \varphi(X)$ <u>il existe</u> $Y \supset X$ <u>tel que</u> $E-Y$ <u>soit fini et</u> $x \notin \varphi(Y)$. En particulier il existe $H \in \mathcal{H}$ tel que $E-H$ soit fini et $X \subset H$, $x \notin H$.

Démonstration : (i) $\Longleftrightarrow$ (ii). Soit X fini $\subset E$. $r(X) = r^{**}(X)$ est équivalent à l'existence de Y fini $\supset X$ tel que $r(X) = \inf\limits_{Z \text{ fini},\, Z \cap Y = \emptyset} (r(X \cup Z) - r(Z))$ i.e. tel que pout tout Z fini $\subset E-Y$ $r(X) \leqslant r(X \cup Z) - r(Z)$ soit $r(X \cup Z) = r(X)+r(Z)$ équivalent à ce que $(X,E-Y)$ est une paire modulaire (cor. 3 de la prop. 2.3).

i $\Longleftrightarrow$ (iii) $M = M^{**}$ est équivalent à $\varphi(X) = \varphi^{**}(X)$ pour tout X fini $\subset E$. Par le lemme 3.1 pour X fini $\subset E$ $\varphi^{**}(X) = \bigcap\limits_{Y \supset X,\, E-Y \text{ fini}} \varphi(Y)$, d'où par le lemme 3.2 l'équivalence avec la condition (iii) de la proposition.

Définition 3. Nous dirons qu'un matroïde vérifiant une des propriétés équivalentes de la proposition 3.4 est discret à l'infini. Le corollaire 3 de la proposition 2.3 fournit l'expression de cette définition en termes de $\mathcal{L}, \mathcal{S}, r$.

En vue de l'application aux matroïdes associés à un multigraphe donnons la proposition suivante :

Proposition 3.5. Soit M un matroïde sur un ensemble E ; (i), (ii), (iii) sont équivalents :

(i) M est discret à l'infini.

(ii) quel que soit X fini $\subset E$, $X \neq \emptyset$, tout ensemble de stigmes de M contenant X et 2 à 2 disjoints en dehors de X est fini.

(iii) quels que soient X fini $\subset E$, $X \neq \emptyset$, et Y fini $\supset X$, tout ensemble de stigmes de M rencontrant X et 2 à 2 disjoints en dehors de Y est fini.

Lemme 3.6. Soit $X \subset E$, $X \neq \emptyset$, tel qu'il existe une infinité de stigmes de M contenant X et 2 à 2 disjoints en dehors de X : alors pour tout $x \in X$ et $Y \supset X-x$ tel que $E-Y$ soit fini on a $x \in \varphi(Y)$.

Démonstration : Soit $\Sigma \subset \mathcal{S}$ un ensemble infini de stigmes de M contenant X et 2 à 2 disjoints en dehors de X et soient $x \in X$ et $Y \supset X-x$ tel que $E-Y$ soit fini. Les $S \cap (E-(Y+x))$ pour $S \in \Sigma$ étant 2à2 disjoints et $E-Y$ étant fini il existe $S_0 \in \Sigma$ tel que $S_0 \cap (E-Y+x)) = \emptyset$: $x \in S_0 \subset Y+x$ i.e. $x \in \varphi(Y)$.

Démonstration de la proposition 3.5

(i)$\Longleftrightarrow$(ii). Soit M discret à l'infini et supposons qu'il existe X fini $\subset E$, $X \neq \emptyset$, et un ensemble infini de stigmes contenant X et 2 à 2 disjoints en dehors de X. Soit $x \in X$, on a $X \in \mathcal{L}$ d'où $x \notin \varphi(X-x)$: le lemme 3.6 et la condition (iii) de la proposition 3.4 sont alors contradictoires.

(ii) $\Longrightarrow$ (iii). Supposons qu'il existe un ensemble $\Sigma \subset \mathcal{S}$ de stigmes recontrant X et 2 à 2 disjoints en dehors de Y : les $S \cap Y$, $S \in \Sigma$, sont non vides, Y étant fini l'un d'eux se retrouve une infinité de fois i.e. il existe $\Sigma' \subset \Sigma$, Σ' infini, et X' fini, $X' \neq \emptyset$, tels que les stigmes de Σ' contiennent X' et soient 2 à 2 disjoints en dehors de X' ce qui contredit (ii).

(iii) $\Longrightarrow$ (i). Soit X fini $\subset E$ et $x \notin \varphi(X)$. L'ensemble des ensembles de stigmes contenant x et 2 à 2 disjoints en dehors de X+x ordonné par inclusion est inductif : par le théorème de Zorn il existe un tel ensemble Σ maximal. Par (iii) Σ est fini : posons $Y = X \cup (E-(\bigcup_{S \in \Sigma} S))$. $Y \supset X$ et $E-Y = (\bigcup_{S \in \Sigma} S)-X$ est fini. Montrons que $x \notin \varphi(Y)$. En effet on a $x \notin Y$, si $x \in \varphi(Y)$ il existe un stigme $S_o \in \mathcal{S}$ tel que $x \in S_o \subset Y+x$. On a $S_o \subset (X+x) \cup (E-(\bigcup_{S \in \Sigma} S))$: S_o est donc disjoint d'avec chacun des $S \in \Sigma$ en dehors de X+x. D'autre part comme $x \notin \varphi(X)$, S_o rencontre $E-(\bigcup_{S \in \Sigma} S)$ i.e. S_o ce qui contredit la maximalité de Σ.

Applications aux matroïdes associés à un multigraphe

Soit G un multigraphe connexe, E l'ensemble de ses arêtes, $\mathbb{C}(G)$ le matroïde sur E dont les stigmes sont les ensembles d'arêtes des cycles élémentaires de G. Les hyperplans de $\mathbb{C}(G)$ sont exactement les complémentaires des cocycles élémentaires de G. Soit $\mathbb{B}(G)$ le matroïde dont les stigmes sont les ensembles d'arêtes des cocycles élémentaires finis de G : on a ainsi $\mathbb{C}^*(G) = \mathbb{B}(G)$.

Proposition 3.7. Une condition nécessaire et suffisante pour que $\mathbb{B}^*(G) = \mathbb{C}(G)$ est qu'entre 2 sommets distincts de G il n'existe jamais une infinité de chaînes 2 à 2 sans arêtes communes.

Démonstration : Supposons qu'il existe 2 sommets distincts de G reliés par une infinité de chaînes 2 à 2 sans arêtes communes : soient $X_i \subset E$, $i \in I$ infini, les ensembles d'arêtes de ces chaînes. Soit $i_o \in I$: $\{X_{i_o} \cup X_i \mid i \in I-i_o\}$ est un ensemble infini de stigmes de $\mathbb{C}(G)$, contenant X_i et 2 à 2 disjoints en dehors de X_{i_o}, d'où $\mathbb{B}^*(G) = \mathfrak{C}^{**}(G) \neq \mathfrak{C}(G)$ par le (ii) de la proposition 3.5

n.b. on peut vérifier directement que chacun des X_i $i \in I$ contient un stigme de $\mathfrak{C}^{**}(G)$ (mais non de $\mathfrak{C}(G)$).

Inversement supposons qu'il existe un ensemble infini Σ de stigmes de $\mathfrak{C}(G)$ contenant un ensemble X fini $\subset E$, $X \neq \emptyset$, et 2 à 2 disjoints en dehors de X. Soit A l'ensemble des sommets de G extrémités des arêtes de X : à tout $S \in \Sigma$ correspond au moins une chaîne **joignant** 2 sommets distincts de A dont les arêtes appartiennent à E-X, les chaînes correspondant à 2 stigmes différents étant sans arêtes communes. A étant fini il existe nécessairement 2 sommets de A reliés par une infinité de telles chaînes 2 à 2 sans arêtes communes.

§4. PROPRIETES DES MATROIDES DISCRETS A L'INFINI

Proposition 4.1. Soit $M(\mathcal{L},\ldots)$ un matroïde sur un ensemble E ; supposons M discret à l'infini, alors :

1). pour tout X fini $\subset E$ $\varphi(X) - \varphi(\emptyset)$ est fini

2). si $E - \varphi(\emptyset)$ est infini E est de rang infini.

Corollaire immédiat du (ii) de la proposition 3.4.

Soit M (resp. M_i $i \in I$) un matroïde sur un ensemble E (resp. E_i $i \in I$). Nous dirons que M est somme directe des matroïdes M_i (et nous écrirons $M = \bigoplus_{i \in I} M_i$) si les E_i $i \in I$ constituent une partition de E et si, de façon équivalente, 1). $\mathcal{L} = \left\{ \bigcup_{i \in I} X_i \mid X_i \in \mathcal{L}_i \text{ pour } i \in I \right\}$,

$$2).\ \mathcal{S} = \bigcup_{i \in I} \mathcal{S}_i,$$

$$3).\ r(X) = \sum_{i \in I} r_i(X \cap E_i) \text{ pour } X \text{ fini} \subset E \text{ ou}$$

$$4).\ \varphi(X) = \bigcup_{i \in I} \varphi_i(X \quad E_i) \text{ pour } X \subset E.$$

Proposition 4.2. Avec les notations précédentes soit $M = \bigoplus_{i \in I} M_i$: condition nécessaire et suffisante pour que M soit discret à l'infini est que chacun des M_i soit discret à l'infini.

Corollaire de la proposition 3.4.

Nous dirons qu'un matroïde est irréductible s'il ne peut être somme directe non triviale. Tout matroïde est somme directe de matroïdes irréductibles.

Proposition 4.3. Un matroïde irréductible discret à l'infini est dénombrable (i.e. l'ensemble sous-jacent est dénombrable).

Démonstration : Soit M un matroïde irréductible discret à l'infini sur un ensemble E et soit $X_o \subset E$, X_o fini non vide. Supposons définie une suite $X_0 \subset X_1 \subset \dots \subset X_n$ de sous-ensembles finis de E telle que pour tout k $0 \leq k < n$ $(X_k, E-X_{k+1})$ soit modulaire. Par la proposition 3.4. il existe X_{n+1} fini $\supset X_n$ tel que $(X_n, E-X_{n+1})$ soit modulaire. On définit ainsi une suite infinie de sous-ensembles finis de E $X_0 \subset X_1 \subset \dots \subset X_n \subset \dots$ telle que $(X_n, E-X_{n+1})$ soit modulaire pour tout $n \in N$. Posons $X = \bigcup_{n \in N}$: $(X, E-X)$ est modulaire (cor. 3 de la prop. 2.3), d'où M étant irréductible, $E=X$ et E est dénombrable.

Soient M un matroïde sur un ensemble E et $F \subset E$. Nous noterons

1). $\overset{\times}{M}_F(\overset{\times}{\mathcal{L}}_F, \dots)$ le sous-matroïde de M sur F (cf. [1] ; $\overset{\times}{M}_F$ est la contraction de M à F dans la terminologie de [4], notée $M \times F$). On a

$$\overset{\times}{\mathcal{S}}_F = \{S | S \in \mathcal{S},\ S \subset F\},$$

$$\overset{\times}{r}_F(X) = r(X) \quad \text{pour } X \text{ fini} \subset F,$$

$$\overset{\times}{\varphi}_F(X) = \varphi(X) \cap F \quad \text{pour } X \subset F, \dots$$

2). $\dot{M}_F(\dot{\mathcal{L}}_F, \dots)$ la contraction de M à F (cf. [1] ; $\dot{M}_F$ est la réduction de M à F dans la terminologie de [4], noté $M.F$). On a

$$\dot{\mathcal{S}}_F = \{\text{éléments minimaux pour l'inclusion des } S \cap F \neq \emptyset,\ S \in \mathcal{S}\},$$

$$\dot{r}_F(X) = \lim_{Y \text{ fini} \subset E-F} (r(X \cup Y) - r(Y)) \quad \text{pour } X \text{ fini} \subset F,$$

$$\dot{\varphi}_F(X) = \varphi(X \cup (E-F)) \cap F \quad \text{pour } X \subset F, \dots$$

Proposition 4.4. Soient M un matroïde sur un ensemble E et $F \subset E$. Supposons M discret à l'infini, alors

1). $\overset{\times}{M}_F$ est discret à l'infini.

2). si $E-F$ est fini $\dot{M}_F$ est discret à l'infini.

Démonstration

1). Soit X fini $\subset F$: il existe Y fini $\supset X$ tel que $(X,E-Y)$ soit modulaire relativement à M (prop. 3.4.). Il est immédiat que $(X,F-(F\cap Y))$ est modulaire relativement à $\overset{*}{M}_F$ (cf. cor. 3 de la prop. 2.3.).

2). Soit X fini $\subset F$: il existe Y fini $\supset X\cup(E-F)$ (fini d'après l'hypothèse) tel que $(X,E-Y)$ soit modulaire relativement à M. On a $E-Y = F-(F\cap Y)$. Soit $S\in\dot{\mathcal{J}}_F$ tel que $S\cap X\neq\emptyset$, $S\cap(F-(F\cap Y))\neq\emptyset$. Il existe $\tilde{S}\in\mathcal{J}$ tel que $\tilde{S}\cap F = S$, d'où $S-(X\cup(F-(F\cap Y))) = \tilde{S}-(X\cup(E-Y))\neq\emptyset$ et par suite $(X,F-(F\cap Y))$ est modulaire relativement à $\dot{M}_F$ (cor. 3 de la prop. 2.3)

Soient E,E' 2 ensembles, Γ une application multivoque de E dans E' — application de E dans $\mathcal{P}(E')$ — telle que Γ^{-1} soit localement finie (Γ^{-1} est l'application multivoque de E' dans E définie pour $x'\in E'$ par $\Gamma^{-1}(x') = \{x\mid x\in E$ tel que $x'\in\Gamma(x)\}$, Γ^{-1} est localement finie si $\Gamma^{-1}(x')$ est fini pour tout $x'\in E'$). Soit $M(\mathcal{L},\ldots)$ un matroïde sur E ; posons $\mathcal{L}' = \{X'\mid X'\subset E'$ tel que quel que soit Y' fini $\subset X'$ il existe $Y\in\mathcal{L}$ et une bijection η de Y sur Y' telle que $\eta(x)\in\Gamma(x)$ pour tout $x\in Y\}$. On sait que $\mathcal{L}'$ est l'ensemble des parties libres d'un matroïde $M'(\mathcal{L}',\ldots)$ sur E'.

Proposition 4.5. Sous les hypothèses précédentes si M est discret à l'infini et Γ localement finie alors M' est discret à l'infini.

Démonstration : Soit A' fini $\subset E'$; posons $A = \Gamma^{-1}(A') = \bigcup_{x'\in A'}\Gamma^{-1}(x')$. A est fini : il existe B fini $\supset A$ tel que $(A,E-B)$ soit modulaire relativement à M (prop. 3.4). Posons $B' = \Gamma(B)$: B' est fini et $B'\supset A'$, montrons que $(A',E'-B')$ est modulaire relativement à M'. Soient X' fini $\subset A'$, Y' fini $\subset E'-B'$ tels que $X',Y'\in\mathcal{L}'$: il existe $X,Y\in\mathcal{L}$ et une bijection ξ de X sur X' (resp. η de Y sur Y') telle que $\xi(x)\in\Gamma(x)$ pour $x\in X$ (resp. $\eta(x)\in\Gamma(x)$ pour $x\in Y$). On a $X\subset A$ et $Y\subset E-B$ d'où $X\cup Y\in\mathcal{L}$ et par suite $X'\cup Y'\in\mathcal{L}'$, les bijections ξ,η se composant en une bijection de $X\cup Y$ sur $X'\cup Y'$ (cf cor. 3 de la prop. 2.3.).

§ 5. RELATIONS ENTRE M ET M*

Proposition 5.1. Soit $M(\mathcal{L},\ldots)$ un matroïde sur un ensemble E,

1). soient $x\in E$, $F\subset E-X$ et posons $F' = E-(F+x)$: les propriétés $x\in\varphi(F)$, $x\in\varphi^*(F')$ s'excluent l'une l'autre. Si F est fini on a $x\in\varphi^*(F)$

2). pour $S \in \mathcal{S}$ on a $E-S \in \mathcal{F}^*$, pour $S' \in \mathcal{S}^*$ $E-S' \in \mathcal{H}$.

Démonstration

1). supposons $x \in \varphi(F)$ et $x \in \varphi^*(F')$: il existe $S \in \mathcal{S}$ tel que $x \in S \subset F+x$ (resp. $S' \in \mathcal{S}^*$ tel que $x \in S' \in F'+x$). On a $|S \cap S'| = 1$, une contradiction. Si F est fini et $x \notin \varphi^*(F)$ on a par la proposition 1.3 $x \in \varphi((E-F)-x) = \varphi(F')$.

2). soit $x \in S$: $x \in \varphi(S-x)$ d'où par 1). $x \notin \varphi^*(E-X)$ i.e. $E-S \in \mathcal{F}^*$. Pour $S' \in \mathcal{S}^*$ $E-S' \in \mathcal{H}$ par la définition 1.

Proposition 5.2. Soit $M(\mathcal{L},\ldots)$ un matroïde sur un ensemble E ; (i), (ii), (iii) sont équivalents :

(i) M est discret à l'infini

(ii) pour tous $x \in E$ et F fini $\subset E-x$ on a $x \in \varphi(F)$ ou $x \in \varphi^*(F')$ où $F' = E-(F+x)$

(iii) pour tout $S \in \mathcal{S}$ on a $E-S \in \mathcal{H}^*$.

Démonstration : (i) $\Longrightarrow$ (ii) , (iii)). Proposition 5.1. appliquée à M^*. (ii) $\Longrightarrow$ (i)) soit X fini $\subset E$ et soit $x \in \varphi^{**}(X)-X$: par la proposition 1.3 $x \notin \varphi^*((E-X)-x)$ d'où par (i)) $x \in \varphi(X)$ i.e. $\varphi^{**}(X) \subset \varphi(X)$. On a ainsi $\varphi(X) = \varphi^{**}(X)$ pour tout X fini $\subset E$ (cf. lemme 3.2.) et donc $M = M^{**}$. (iii) $\Longrightarrow$ (i) on montre que (iii) entraîne la condition (ii) de la proposition 3.5 : supposons qu'il existe X fini $\subset E$, $X \neq \emptyset$, et un ensemble infini Σ de stigmes de M contenant X et 2 à 2 disjoints en dehors de X. Soit $S \in \Sigma$; $S \supsetneq X$, soient $x \in X$ et $y \in S-X$. Par (iii) $E-S \in \mathcal{H}^*$ d'où $x \in \varphi^*(E-S)+y)$). $x \notin (E-S)+y$: il existe Y' fini $\subset (E-S)+y$ tel que $x \notin \varphi(E-Y')-x)$). Posons $Y = (E-Y')-x$: on a $Y \supset X-x$, $E-Y$ fini et $x \notin \varphi(Y)$ contredisant le lemme 3.6.

Lemme 5.3. Soit $M(\mathcal{L},\ldots)$ un matroïde sur un ensemble E ; les propositions suivantes sont équivalentes :

(i) $\mathcal{B}^* = \{E-B | B \in \mathcal{B}\}$

(ii) quel que soit $X \subset E$ et $x \notin \varphi(X)$ il existe $Y \supset X$ tel que $E-Y$ soit fini et $x \notin \varphi(Y)$

(iii) pour tous $x \in E$, $F \subset E-x$ on a $x \in \varphi(F)$ ou $x \in \varphi^*(F')$ où $F' = E-(F+x)$

(iv) <u>pour tout</u> $H \in \mathcal{H}$ <u>on a</u> $E-H \in \mathcal{S}^*$

(v) <u>pour tout</u> $H' \in \mathcal{H}^*$ <u>on a</u> $E-H \in \mathcal{S}$

(vi) <u>les</u> $E-H \in \mathcal{H}$ <u>sont finis</u>

<u>Démonstration</u> : On montre que (iii) $\Longrightarrow$ (i) $\longrightarrow$ (ii) $\Longrightarrow$ (vi) $\Longrightarrow$ (iv) (iii) puis (vi) $\Longrightarrow$ (v) $\Longrightarrow$ (iii).

(iii) $\longrightarrow$ (i) . soit $B \in \mathcal{B}$: $E-B \in \mathcal{L}^*$ (prop. 1.3). Soit $x \in B$: on a $x \notin \varphi(B-x)$ d'où par (iii) $x \in \varphi^*(E-B)$ i.e. $E = \varphi^*(E-B)$, $E-B \in \mathcal{B}^*$. Inversement soit $B' \in \mathcal{B}^*$: $E-B' \in \mathcal{L}^{**} \subset \mathcal{L}$ (lemme 3.2). Soit $x \in B'$: $x \notin \varphi^*(B'-x)$ d'où par (iii) $x \in \varphi(E-B')$ i.e. $B' \in \mathcal{B}$.

(i) $\Longrightarrow$ (ii). soient $X \subset E$ et $x \notin \varphi(X)$. Soit X_o une base de X : $X_o + x \in \mathcal{L}$ et soit $B \in \mathcal{B}$, $B \supset X_o+x$. Par (ii) $E-B \in \mathcal{B}^*$ d'où $x \in \varphi^*(E-B)$ et il existe Y' fini $\subset E-B$ tel que $x \notin \varphi((E-Y')-x))$ (prop. 1.3) Posons $Y = \varphi((E-Y')-x)$: $(E-Y')-x \supset B-x \supset X_o$ d'où $Y \supset \varphi(X_o) = \varphi(X) \supset X$ et $E-Y \subset Y' + x$ est fini.

(ii) $\Longrightarrow$ (vi) soit $H \in \mathcal{H}$: il existe $x \in E-H$ d'où par (ii) l'existence de $Y \supset H$ tel que $E-Y$ soit fini et $x \notin \varphi(Y)$. Nécessairement $Y=H$ et $E-H$ est fini.

(vi) $\longrightarrow$ (iv) évident.

(iv) $\Longrightarrow$ (iii) supposons $x \notin \varphi(F)$: il existe $H \in \mathcal{H}$ tel que $x \notin H$ et $H \supset F$. Par (iv) $E-H \in \mathcal{S}^*$ d'où $x \in E-H \subset F'+x$ i.e. $x \in \varphi^*(F')$.

(vi) $\Longrightarrow$ (v) par ce qui précède (vi) entraîne (ii) : en particulier M est discret à l'infini, de même (vi) entraîne (i) d'où $\mathcal{B}^* = \{E-B \mid B \in \mathcal{B} = \mathcal{B}^{**}\}$ i.e. $\mathcal{B}^{**} = \{E-B' \mid B' \in \mathcal{B}^*\}$. M* vérifie donc (i) et par suite (iv) qui s'écrit $H \in \mathcal{H}^*$ entraîne $E-H \in \mathcal{S}^{**} = \mathcal{S}$.

(v) $\Longrightarrow$ (iii) supposons que $x \notin \varphi^*(F')$: il existe $H' \in \mathcal{H}^*$ tel que $x \notin H'$, $H' \supset F'$. par (v) $E-H' \in \mathcal{S}$ d'où $x \in E-H' \subset F+x$ i.e. $x \in \varphi(F)$.

Aucune des propriétés équivalentes du lemme 5.3. toutes vérifiées dans le cas où E est fini (cf propriété (vi)) n'est vérifiée lorsque M infini n'est pas somme directe de matroïdes finis. En effet :

<u>Proposition 5.4.</u> <u>Soit</u> M($\mathcal{L}$,...) <u>un matroïde irréductible sur un ensemble</u> E <u>infini : il existe</u> $H \in \mathcal{H}$ <u>tel que</u> $E-H$ <u>soit infini.</u>

Pour la démonstration introduisons la définition suivante : soient $X \in \mathcal{L}$, $S \in \mathcal{S}$, nous dirons que S <u>est un pont relativement à</u> X si 1). $X \cap S \neq \emptyset$ et 2). il existe unebase A de $S \cup X$ telle que $X \subset A$ et $|S-A| = 1$.

Lemme 5.5.

(i) soit $S \in \mathcal{T}$ un pont relativement à $X \in \mathcal{L}$: pour tout $x \in S$ $(S \cup X)-x \in \mathcal{L}$ et par suite S est le seul stigme de $\mathcal{T}$ contenu dans $S \cup X$.

(ii) soient $X \in \mathcal{L}$, $S \in \mathcal{T}$ tels que $S \cap X \neq \emptyset$ et $x \in S-X$: il existe un pont $T \in \mathcal{T}$ relativement à X tel que $x \in T \subset S \cup X$

(iii) soient $S \in \mathcal{T}$ un pont relativement à $X \in \mathcal{L}$ et $x \in E-(S \cup X)$ tels qu'il existe un stigme $T \in \mathcal{T}$ contenu dans $X+x$: pour tout $y \in S \cap T$ il existe un stigme unique $R \in \mathcal{T}$ contenu dans $(S \cup (X+x))-y$ et on a $x \in R$ et $S-X \subset R$.

Démonstration

(i) soit A une base de $S \cup X$ telle que $X \subset A = (S \cup X)-a$ où $a \in S-X$. Si $(S \cup X)-x \notin \mathcal{L}$ il existe $T \in \mathcal{T}$ tel que $T \subset (S \cup x)-x$: nécessairement $x \neq a$, $a \in T$ et $T \neq S$, par élimination de a entre S et T il existe $S' \in \mathcal{T}$ $S' \subset (S \cup T)-a \subset A$ une contradiction.

(ii) soit A une base de $(S \cup X)-x$ telle que $A \supset X$: on a $\varphi(A) \supset S-x$ d'où $x \in \varphi(A)$ et il existe un stigme $T \in \mathcal{T}$ tel que $x \in T \subset A+x$ $S \cup X$. $T \cap X \neq \emptyset$ sinon $T \subset S-X \subsetneq S$ et $A \cap (T \cup X) = X \cup (A \cap T) = (X \cup T)-x$ est une base de $X \cup T$.

(iii) soit $z \in S-X \subset S-T$: il existe $R \in \mathcal{T}$ tel que $z \in R \subset (S \cup T)-y \subset (S \cup (X+x))-y$. Par (i) tout stigme contenu dans $(S \cup (X+x))-y$ contient x : il existe par suite un seul stigme contenu dans $(S \cup (X+x))-y$ d'où l'unicité de R et $x \in R$, $S-X \subset R$.

Démonstration de la proposition 5.4

Soit $\{x_o, x_1, \dots\}$ un sous-ensemble infini de E indexé par $\mathbb{N}$. M étant irréductible pour $i \geq 1$ il existe $R_i^o \in \mathcal{T}$ contenant $x_o = e_o$ et x_i. Posons $R_1 = R_1^o$ et soit $J_o = \{i \in \mathbb{N} \mid x_i \notin R_1\}$: J_o est infini. Pour $i \in J_o$ il existe $T_i^o \in \mathcal{T}$ tel que $x_i \in T_i^o \subset (R_1 \cup R_i^o)-e_o$. $T_i^o \cap (R_1-e_o) \neq \emptyset$: par le (ii) du lemme 5.5. il existe un pont $S_i^1 \in \mathcal{T}$ relativement à R_1-e_o tel que $x_i \in S_i^1 \subset T_i^o \cup (R_1-e_o)$. J_o est infini, R_1-e_o est fini et $S_i^1 \cap (R_1-e_o) \neq \emptyset$ pour $i \in J_o$: il existe $e_1 \in R_1-e_o$ et I_1 infini $\subset J_o$ tel que $e_1 \in S_i^1$ pour tout $i \in I_1$. On pose $A_1 = R_1-e_o$ et $\Sigma_1 = (S_i^1)_{i \in I_1}$. Soit $n \geq 1$: pour $k = 1,2,\dots,n$ supposons définis A_k fini $\in \mathcal{L}$, $e_k \in E$, $I_k \subset \mathbb{N}$ et $\Sigma_k = (S_i^k)_{i \in I_k}$ où $S_i^k \in \mathcal{T}$ avec les propriétés suivantes :

i) $e_o \in A_1 \subset A_2 \subset \dots \subset A_n$. Les e_k $k = 1,2,\dots,n$ sont 2 à 2 distincts et pour $1 \leq k \leq n$ il existe $R_k \in \mathcal{T}$ tel que $e_o \in R_k \subset A_k + e_k$.

ii) I_k est infini et pour tout $i \in I_k$ $x_i \in S_i^k - (A_k+e_k)$, $e_k \in S_i^k$, $e_o, e_1,\dots,e_{k-1} \notin S_i^k$, S_i^k est un pont relativement à $(A_k+e_k)-e_o$.

Soit alors $i \in I_n$: par le (iii) du lemme 5.5 (avec $S = S_i^n$, $X = (A_n+e_n)-e_o$, $x = e_o$ et $T = R_n$) $(S_i^n \cup A_n)-e_n$ contient un stigme unique soit R_i^n et on a $e_o \in R_i^n$ et $S_i^n - (A_n+e_n) \subset R_i^n$ (en particulier $x_i \in R_i^n$). Soit $i_n \in I_n$ quelconque (par exemple le plus petit), posons $S_n = S_{i_n}^n$, $R_{n+1} = R_{i_n}^n$ et soit $J_n = \{i \in I_n \mid x_i \notin S_n\}$. Pour $i \in J_n$ il existe $T_i^n \in \mathcal{T}$ tel que $x_i \in T_i^n \subset (R_{n+1} \cup R_i^n)-e_o \subset (A_n \cup S_n \cup S_i^n) - (e_o+e_n)$. $T_i^n \cap ((A_n \cup S_n) - (e_o+e_n)) \neq \emptyset$: par le (ii) du lemme 5.5 il existe un pont $S^{n+1} \in \mathcal{T}$ relativement à $(A_n \cup S_n) - (e_o+e_n)$ tel que $x_i \in S_i^{n+1} \subset T_i^n \cup ((A_n \cup S_n) - (e_o+e_n))$ $\subset (A_n \cup S_n \cup S_i^n) - (e_o+e_n)$. $S_i^{n+1} \cap (S_n-(A_n+e_n)) \neq \emptyset$: en effet sinon on aurait $S_i^{n+1} \subset (A_n \cup S_i^n) - (e_o+e_n)$ ce qui contredirait l'unicité du stigme contenu dans $(A_n \cup S_i^n)-e_n$ (car $R_i^n \neq S_i^{n+1}$). J_n est infini, $S_n-(A_n+e_n)$ est fini et $S_i^{n+1} \cap (S_n-(A_n+e_n)) \neq \emptyset$ pour $i \in J_n$: il existe $e_{n+1} \in S_n-(A_n+e_n)$ et I_{n+1} infini $\subset J_n$ tel que $e_{n+1} \in S_i^{n+1}$ pour $i \in I_{n+1}$. On pose alors $A_{n+1} = (A_n \cup S_n) - (e_n+e_{n+1})$ et $\Sigma_{n+1} = (S_i^{n+1})_{i \in I_{n+1}}$ ce qui achève la construction à l'étape $n+1$. La vérification des hypothèses de récurrence est immédiate.

Posons alors $A = \bigcup_{n \in \mathbb{N}} A_n$: $A \in \mathcal{L}$ et il existe $B \in \mathcal{B}$ tel que $A \subset B$. $e_o \in B$ soit $H = \varphi(B-e_o)$ on a $e_o, e_1, e_2, \dots \notin H$ établissant la proposition.

- Application aux graphes

Proposition 5.6. Un multigraphe 2-connexe dont tous les cocycles élémentaires sont finis est fini.

Bibliographie

[1] H.H. CRAPO, G.C. ROTA: Combinatorial Geometrics (1968)

[2] M. LAS VERGNAS: Sur la dualité en théorie des matroïdes, Comptes-Rendus Acad. Sci., Paris 270 (1970), 804-806

[3] P. Robert: Sur l'axiomatique des systèmes générateurs, des rangs, ... Thèse Rennes 1966. Bull. Soc. Math. France, mémoire n° 14 (1968)

[4] W.T. Tutte: Lectures on matroïds, J.Res.Nat.Bur.Stand. 69 B (1965) 1-47

A CHARACTERIZATION OF TRANSVERSAL INDEPENDENCE SPACES

J. H. Mason

Abstract

Given independence spaces on the sets E and Y, necessary and sufficient conditions are obtained for the existence of a relation $R \subseteq E \times Y$ such that R and the independence space on Y induce the independence space on E. As a special case transversal independence spaces are characterized by their rank functions.

1. Introduction

An independence space on a set E, is a pair $(E, \mathcal{C})$ where $\mathcal{C}$ is a collection of finite subsets of E called circuits satisfying the following condition: If C_1 and C_2 are distinct members of $\mathcal{C}$, then

(i) C_1 is not contained in C_2

(ii) if $e \in C_1 \cap C_2$ and $f \in C_1 - C_2$ then there exists a circuit C

$$C \subseteq C_1 \cup C_2 - \{e\} \text{ with } f \in C.$$

If we denote by $\mathcal{C}/F$ the members of $\mathcal{C}$ contained in the subset F of E, then clearly $(F, \mathcal{C}/F)$ is an independence space on F called the restriction to F. A subset of E is called dependent if it contains a circuit, and independent otherwise, and a maximal independent set is called a basis.

The terms matroid and combinatorial pre-geometry are generally used in the case that E is a finite set, 'matroid' having been introduced by Whitney [8] as an abstraction of the concept of linear independence of the rows of a matrix, and combinatorial pre-geometry having been introduced by Crapo-Rota [1] in the context of geometric lattices. Both of these papers contain several alternative axiom schemes. We will use the term independence space (shortened to 'space') to infer that E may be finite or infinite.

The important property of a matroid is that all bases have the same cardinality, and the finiteness of circuits implies that all bases of an independence space also have the same cardinality. Thus we can define the rank of a subset A of E to be the cardinality of a basis of $(A, \mathcal{C}/A)$. The span of the set A is the union of A with the set of all e in E for which there exists a circuit C contained in $A \cup \{e\}$ and containing e. Finally, we shall denote by $\mathcal{F}(E)$ the collection of finite unions of circuits of $(E, \mathcal{C})$. Tutte [7] calls these flats, but since the term is now used for a different purpose (Crapo-Rota [1]), we shall call them the fully dependent subsets of E, since if f belongs to a member F of $\mathcal{F}(E)$, there is a circuit containing f and contained in F.

We are interested in inducing independence spaces through a binary relation. If $R \subseteq E \times Y$, and if F is a subset of E, and X is a subset of Y, then we say that F and X are linked in R if there is a bijection of F onto X which is a subset of R. Now if $(Y, \mathcal{C}_Y)$ is an independence space, then we call R locally right rank finite (LRRF) if the set

$$R(e) = \{y \ : \ (e,y) \in R\}$$

has finite rank for every e in E. R is locally right finite (LRF) if R(e) has finite cardinality for every e in E. Clearly, if $\mathcal{C}_Y = \emptyset$ then LRRF and LRF are the same. Finally, for ease in notation, we put R(A) equal to union of the R(a) for $a \in A$.

It is well known (Perfect [5]) that $(Y, \mathcal{C}_Y)$ and R induce an independence space on E if R is LRF. The proof, which we will not give here, is based on the following theorem of Rado, which is itself based on the Rado selection principle.

Theorem 1 .1 (Rado [6]) If $R \subseteq E \times Y$ is a LRF relation and $(Y, \mathcal{C}_Y)$ is an independence space, then E can be linked in R to an independent subset of Y if and only if

$$(*) \qquad \text{rk } R(F) \geq \text{card } F$$

for each finite subset of E.

We will need an apparently stronger version of this theorem, which is actually a corollary.

Corollary 1 .1 If $(Y, \mathcal{C}_Y)$ is an independence space on Y and $R \subseteq E \times Y$ is a LRRF relation then E can be linked in R to an independent subset of Y if and only if condition (*) holds.

Proof: For each e in E, let B_e be a basis of R(e) and put $R^* = \{(e,b) : b \in B_e\}$. Since R is LRRF, R* is LRF. Since $R^* \subseteq R$, the result follows from theorem 1.

An immediate consequence of Corollary 1, using the same proof as in Perfect [5] , we get the following theorem.

Theorem 1.2 If $(Y, \mathcal{E}_Y)$ is an independence space on Y and $R \subseteq E \times Y$ is LRRF, then an independence space $(E, \mathcal{E}_E)$ is induced on E where independent sets are those subsets of E linked in R to independent subsets of Y.

It is important to note that no great generality is achieved, for in the light of the proof of the corollary, any LRRF relation can be replaced by a LRF relation which induces the same space E. However it will facilitate matters to have this "more general" condition on R, and we will assume throughout that all relations are LRRF.

In the special case that $\mathcal{E}_Y = \emptyset$, then the space induced on E is called a transversal space (Edmonds [2]) since the independent subsets of E are the partial transversals of the family (R(e) : $e \in E$) of finite subsets of E.

The purpose of this paper is to give a characterization of transversal spaces in terms of the rank function, and more generally, to provide necessary and sufficient conditions that, given the space $(E, \mathcal{E}_E)$ and $(Y, \mathcal{E}_Y)$, there exists a relation $R \subseteq E \times Y$ which will induce the space on E from the space on Y.

2. Preliminary Theorems

Before we tackle the main theorem, we need two straightforward results. The first shows how the restriction of an induced space is induced, and the second shows how to construct a canonical relation to induce one space from another, which is in a sense maximal.

Theorem 2.1 (Restriction Theorem) Let $(E, \mathcal{C}_E)$ be induced by $R \subseteq E \times Y$ and $(Y, \mathcal{C}_Y)$. Let B be a basis of $(E, \mathcal{C}_E)$ which is linked to an independent set X contained in Y. Then $(E, \mathcal{C}_E)$ is induced by $R^* = R \cap (E \times \text{Span } X)$.

Proof If e is a member of E - B then R(e) is contained in Span X since otherwise $B \cup \{e\}$ would be independent in $(E, \mathcal{C}_E)$. Let $(E, \mathcal{C}_0)$ be the space induced on E by R^* and $(Y, \mathcal{C}_Y)$; clearly every independent set in $(E, \mathcal{C}_0)$ is independent in $(E, \mathcal{C}_E)$, so we need only show that if a set A is independent in $(E, \mathcal{C}_E)$ it is independent in $(E, \mathcal{C}_0)$. To this end, assume not, and let A be minimal with respect to being dependent in $(E, \mathcal{C}_0)$ but independent in $(E, \mathcal{C}_E)$. Thus A is finite, and not contained in B. There must therefore exist b in $A \cap B$ with $(b,y) \in R$ but $y \notin$ Span X for some y. Now for each a in A - B there exists a circuit C_a in $\mathcal{C}_E$ with C_a contained in $B \cup \{a\}$. Since each C_a is dependent in $(E, \mathcal{C}_0)$, it is easy to show that

$$A \subseteq \bigcup C_a \quad [a \in A - B]$$

whence for some a in A - B, b is a member of C_a. One easily checks then that $(B - \{b\}) \cup \{a\}$ is a basis of $(E, \mathcal{C}_E)$ and of $(E, \mathcal{C}_0)$ and hence is linked in R to an independent subset S of Span X. Then $B \cup \{a\}$ would be linked to the independent subset $S \cup \{y\}$ of $(Y, \mathcal{C}_Y)$ contradicting the maximality of B.

Corollary 2.1. Let $R \subseteq E \times Y$ and $(Y, \mathcal{C}_Y)$ induce $(E, \mathcal{C}_E)$, and let F be a subset of E. If B is a basis of $(F, \mathcal{C}_E/F)$, and B is linked to an independent subset Z of Y, then $(F, \mathcal{C}_E\, F)$ is induced by $R^* = R \cap (F \times \text{Span } Z)$.

Proof: Clearly $R \cap (F \times Y)$ induces $(F, \mathcal{C}_E\, F)$, and the result follows from theorem 2.1.

Since in general there are many LRRF relations $R \subseteq E \times Y$ which, with $(Y, \mathcal{C}_Y)$, will induce the same space on E, it will be convenient to have a canonical one to refer to. The purpose of the next theorem is to construct a relation containing a given relation, which is maximal (by inclusion) with respect to inducing the same space on E. Now clearly it is not possible to do this with a LRRF relation if there exists an e in E contained in no circuit, for then e belongs to every basis and so could be related to every member of Y. However it is also clear that as long as rk Y $\geq$ rk E, we may ignore any such elements when we try to construct a relation $R \subseteq E \times Y$ to induce the space $(E, \mathcal{C}_E)$, since we can make use of the restriction theorem. Thus it will cause no loss of generality to assume that every element of E belongs to a circuit, as long as we keep rk Y $\geq$ rk E.

Theorem 2.2 (Maximal Relations) If $R \subseteq E \times Y$ and $(Y, \mathcal{C}_Y)$ induce $(E, \mathcal{C}_E)$, then for each e in E, put

$$R^*(e) = \bigcap \text{Span } R(C) \left[e \in C \in \mathcal{C}_E\right]$$

Then $R^*(e)$ and $(Y, \mathcal{C}_Y)$ induce $(E, \mathcal{C}_E)$, $R \subseteq R^*$, and R^* is maximal such.

Proof: The rank of $R^*(e)$ is finite for each e in E since for some C in $\mathcal{C}_E$, e is a member of C and so $R^*(e)$ is contained in Span R(C) which has finite rank. Thus R^* is LRRF. It is clear that R(e) is contained in $R^*(e)$, and so we need only show that no new independent sets are introduced by R^*. Therefore we need only show that if C_0 is a circuit in $\mathcal{C}_E$, C_0 is dependent as induced by R^*. This will certainly be the case if rk $R(c_0)$ = rk $R^*(C_0)$. But if e is a member of C_0, $R^*(e)$ is contained in Span C_0. Therefore

$$\text{Span } R^*(C_0) \subseteq \text{Span } R(C_0) \subseteq \text{Span } R^*(C_0)$$

and so rk $R(C_0)$ = rk $R^*(C_0)$. The maximality of R^* follows easily.

3. Characterization of Transversal Spaces

Theorem 3.1. If $(E,\mathcal{C}_E)$ and $(Y,\mathcal{C}_Y)$ are independence spaces with rk Y $\geq$ rk E, then there exists a LRRF relation $R \subseteq E \times Y$ which with $(Y,\mathcal{C}_Y)$ induces $(E,\mathcal{C}_E)$ if and only if there exists a map

$\sigma : \mathcal{F}(E) \to \mathcal{F}(Y)$ with the following properties

(Ai) $\quad rk^{E}(F) = rk^{Y}(\sigma(F))$ for all F in $\mathcal{F}(E)$

(Aii) $\quad \sigma(F_1 \cup F_2) = \sigma(F_1) \cup \sigma(F_2)$ for all F_1, F_2 in $\mathcal{F}(E)$

(Aiii) $\quad rk^{E}(A) \leq rk^{Y}\sigma(A)$ for every finite subset A of E

Where $\sigma(A) = \bigcup_{a \in A} \left(\bigcap_{a \in C \in \mathcal{C}_E} \text{Span}\, \sigma(C) \right)$

In the light of (Aii), σ is determined by its action on the circuits $\mathcal{C}_E$.

<u>Proof</u>: If we are given R, then we may assume it to be maximal in the sense of Theorem 2.2. Taking $\sigma(F) = R(F)$ for all F in $\mathcal{F}(E)$, (Aii) is trivial and (Aiii) is a restatement of the fact that $rk^{E} A \leq rk^{Y} R(A)$. To establish (Ai), let B be a basis of $(F, \mathcal{C}_E/F)$. B is linked in R to an independent subset Z of Y, and so by maximality of B, R (F - B) is contained in Span Z. By an argument similar to that of theorem 1.1, we find R (B) is contained in Span Z and so $rk^{E}(F) = rk^{Y} R(F)$ which is (Ai).

Conversely, if we are given the map σ, put

$$R(E) = \bigcap \text{Span}\, \sigma(C) \qquad [e \in C \in \mathcal{C}_E].$$

Now since for each e we can assume there is a circuit C containing e, R (e) is contained in Span $\sigma(C)$ which has finite rank, so R is LRRF. Then R and $(Y, \mathcal{C}_Y)$ induce a space $(E, \mathcal{C}_0)$ on E. To show that $\mathcal{C}_E = \mathcal{C}_0$, we first let C be a member of $\mathcal{C}_E$. C is dependent in $(E, \mathcal{C}_0)$ since R (C) is contained in Span $\sigma(C)$ and so

$$rk^{Y} R(C) \leq rk^{Y} \sigma(C) = rk^{E} C = \text{Card } C - 1$$

Secondly, if A_0 is independent in $(E, \mathcal{C}_E)$, then for any finite subset A of A_0 we have by (Aiii) that $rk^{Y} R(A) \geq rk^{E} A = \text{Card } A$. By corollary 1.1, A is independent in $(E, \mathcal{C})$ and hence so is A_0 and the proof is complete.

The three conditions contained above are very messy, but in the case of $\mathcal{C}_Y = \emptyset$ they can be considerably neatened.

Corollary 3.1 $(E, \mathcal{C})$ is a transversal space if and only if there exists a map $\sigma : \mathcal{F}(E) \rightarrow P\omega(Y)$ such that

(Bi) $\quad rk^E(F) = \text{Card}\ (F)$ for all F in $\mathcal{F}(E)$

(Bii) $\quad \sigma(F_1 \cup F_2) = \sigma(F_1) \cup \sigma(F_2)$ for all F_1, F_2 in $\mathcal{F}(E)$

(Biii) $\quad rk^E(A) \leq \text{Card} \bigcap \sigma(F) \quad [A \subseteq F \in \mathcal{F}(E)]$ for any finite subset A of E.

Proof: One readily checks that (Aiii) and (Biii) are equivalent when $\mathcal{C}_Y = \emptyset$.

The conditions for the existence of a map σ satisfying (Bi) and (Bii) are just the conditions that the set of numbers $\{rk\ F : F \in \mathcal{F}(E)\}$ must satisfy in order that they be the cardinalities of the unions of a family of finite sets, and these are treated in [3]. When applied to this situation they give the following theorem.

Theorem 3.2 $(E, \mathcal{C})$ is a transversal space if and only if for any $F_1, \ldots F_n$ in $\mathcal{F}(E)$,

$$\sum (-1)^{1 + \text{Card}\ I}\ rk\ F(I)\ [I \subseteq N] \geq rk \bigcap F_i\ [i \in N]$$

where $N = \{1, \ldots, n\}$ and $F(I) = \bigcup F_i\ [i \in I]$

Proof: In the presence of (Bi) and (Bii), (Biii) is equivalent to the stated condition. By the analysis in [3], the map σ exists satisfying (Bi) and (Bii) if and only if the stated condition holds.

Remark 1: It appears to be exceedingly difficult to find a condition such as in Theorem 3.2 to deal with the more general condition (Aiii).

Remark 2: If $(E, \mathcal{C})$ is a transversal space, and E is a finite set, then the analysis in [3] shows that $(E, \mathcal{C})$ is induced by a unique maximal relation up to a permutation of the elements of Y.

The condition in Theorem 3.2 gives rise to a sufficient but not necessary condition for $(E, \mathcal{C})$ to be non-transversal.

Corollary 3.2 If $(E, \mathcal{C})$ has rank k, and if it contains more than k k-element circuits whose pair-wise unions contain bases, then $(E, \mathcal{C})$ is not transversal.

Proof: Let $C_1, \ldots, C_n$ be distinct k-element circuits whose pair-wise unions contain bases. By Theorem 3.2, if $(E, \mathcal{C})$ were transversal, we would have

$$\binom{n}{1}(k-1) - \binom{n}{2} k + \binom{n}{3} k - \ldots \geq 0$$

i.e. $k-n \geq 0$ i.e. $n \leq k$. Thus if $n > k$, $(E, \mathcal{C})$ cannot be transversal.

Bibliography

[1] H.H. Crapo & G.-C. Rota, Combinatorial Geometrics, privately printed, Dec, 1969

[2] J. Edmonds, Systems of Distinct Representatives and Linear Algebra, J. Res. Nat. Bur. Standards, Sec. B. 69 (1965), 147-153.

[3] J.H. Mason, Representations of Independence Spaces, Ph.D.Thesis, Univ. of Wisconsin (1969)

[4] L.Mirsky & H.Perfect, Applications of the Notion of Independence to Problems of Combinatorial Analysis, J.Combinatorial. Theory 2 (1967) 325-57.

[5] H.Perfect, Independence Spaces and Combinatorial Problems, Proc. Lond. Math. Soc. (3) 19 (1969), 17-30

[6] R. Rado, Axiomatic Treatment of Rank in Infinite Sets, Canadian Math. J. 1 (1949) 337-343.

[7] W.T. Tutte, Introduction to the Theory of Matroids, Rand. Report, R448 PR (1966), 1-98.

[8] H.Whitney, On the Abstract Properties of Linear Dependence, Amer. J. Math. 57 (1935), 509-533.

MATROIDS AND BLOCK DESIGNS

D.J.A. Welsh

1. Introduction

My interest in the relation between matroids and block designs originated in a joint paper with J.A. Bondy [1] where we obtain the following minimal presentation of a transversal matroid.

Theorem 1. If $\underline{M}$ is a transversal matroid on S, of rank r, there exist distinct cocircuits $C_1^*, \ldots, C_r^*$, such that $\underline{M}$ is the family of partial transversals of the family $(C_i^*: 1 \leq i \leq r)$.

This theorem leads to the definition of an intersection operator Λ for matroids and we prove that with respect to this operator every matroid can be expressed as the intersection of transversal matroids.

Theorem 2. If $\beta(\underline{M})$ denotes the set of bases of $\underline{M}$, let

$$\beta\left(\bigwedge_{i=1}^{k} \underline{M}_i\right) = \bigcap_{i=1}^{k} \beta(M_i).$$

Then if $\underline{M}$ is any matroid there exist transversal matroids $\underline{M}_i$ $(1 \leq i \leq p)$ such that

$$\beta(\underline{M}) = \beta\left(\bigwedge_{i=1}^{r} \underline{M}_i\right)$$

The crux of the proof of Theorem 2 is the following theorem which historically was the origin of these theorems.

Theorem 3. _If_ B^* _is any cobase of a matroid_ M _and_ $C^* \equiv (C_i^*: 1 \leq i \leq r)$ _are the fundamental cocircuits determined by_ B^*, _then every base of_ M _is a transversal of_ C^*.

Theorem 1 leads to the upper bound $n\ 2^{n^2}$ for the number of non-isomorphic transversal matroids on a set of n elements. We conjectured that this could be reduced to $n\ 2^{n^2/2}$, and this would follow if we could have shown that any transversal matroid of rank r on a set of n elements had a presentation of the form $(A_1, A_2, \ldots, A_r)$ where

$$A_i \subseteq S - \{x_1, x_2, \ldots, x_{i-1}\} \qquad (1 \leq i \leq r),$$

$$S = \{x_1, \ldots, x_r\}.$$

We eventually found a family of counterexamples to this.

Let $S = \{1, \ldots, 4n\}$. Let A_i $(1 \leq i \leq n)$ be any family of subsets of S with the following properties

$$|A_i \cap A_j| = n.$$

$$|A_i| = 2n.$$

The transversal matroid defined by the family

$$(A_1, \ldots, A_n, S-A_1, \ldots, S-A_n)$$

is then seen to be a counterexample to the above conjecture.

It is easily seen that matroids of this type exist whenever there exists a $(4n-1, 2n-1, n-1)$ symmetric design. Moreover each of these matroids has the property of being _identically self-dual_ (ISD). That is every circuit is a cocircuit and vice versa.

Various properties of ISD matroids have been proved in [1]. In particular we show that whereas there is no connected graphic

ISD matroid we do have

Theorem 4. For any n for which a (4n-1, 2n-1, n-1) design exists, there exists a connected, transversal, ISD matroid on a set of 4n elements.

Although there are no graphic ISD connected matroids we also prove

Theorem 5. For any set of cardinality 2n there exists a non-trivial, connected, binary, ISD matroid on S provided there exists a (v,k,λ) design with parameters $v = n$, k odd and λ even.

These two theorems give disjoint sets of matroids because of the following result proved in [3]

Theorem 6. There exist no binary transversal matroids except those which are graphic and transversal.

In [6] we show that for any group G (finite or infinite) there exists a matroid M whose automorphism group is abstractly isomorphic to the group G. Calling M s-transitive if its automorphism group is s-transitive we find that this leads easily to a characterisation of all matroids with $A(\underline{M}) \cong S_n$ and $A(\underline{M}) \cong A_n$ (the alternative group). Moreover there exist interesting matroids with high transitivity. For example, let S(d,k,n) denote the Steiner system consisting of a set S of cardinality n, a family of blocks B_i such that $|B_i| = k$ and such that every d-subset of S belongs to a unique block. Recognising these as paving geometries (see [2]) we see that the blocks of S(d,k,n) are the hyperplanes of a matroid M of rank d+1. In particular from S(5,6,12) we get a matroid of rank 6 on a set of 12 elements with automorphism group the

5-transitive Mathieu group $\underline{M}_{12}$. More recently Piff [11] has used 'design ideas' to construct matroids on n elements having the cyclic group C_n as automorphism group for all but a finite set of values of n. These show that the class of "matroid groups" is much larger than the set of "graphic groups".

Another use of designs is in the construction of equicardinal matroids. This theory is due to U.S.R. Murty [9], [10] and he has characterised those binary matroids in which every hyperplane has the same cardinality. Essentially he shows them to be just those block designs derivable from geometries over GF(2).

Thus the problem arises, for what balanced incomplete block designs do the blocks of the design form the hyperplanes of a matroid? Our matroid terminology will be standard as in [7] and the design terminology will be that of Hall's book [5].

2. Definitions

Throughout $\underline{D}$ will denote a block design on the ground set S and $\underset{\sim}{D}$ (b,v,r,k,λ) denotes a balanced incomplete block design (BIBD) with parameter set (b,v,r,k,λ). When the design or parameter set is symmetric we modify this notation accordingly. Call a design $\underline{D}$ a matroid design or matroidal if (a) it has distinct blocks (regarded as sets) and (b) its blocks satisfy the following hyperplane axioms for a matroid;

Hyperplane Axioms If H_1, H_2, are distinct hyperplanes and x is an element not belonging to $H_1 \cup H_2$ there exists a hyperplane H_3 such that

$$\{x\} \cup (H_1 \cap H_2) \subseteq H_3 .$$

We call the parameter set (b,v,r,k,λ) matroidal if there exists some matroid design with these parameters. The parameters (b,v,r,k,λ) are strongly matroidal if every design with distinct blocks and with these parameters is matroidal. A parameter set is non-matroidal if no design with these parameters is matroidal.

Throughout when discussing designs we shall be ignoring the trivial degenerate cases where the blocks of the design are all k-sets for some integer k. We shall also only consider those parameter sets for which designs are known to exist.

3. Symmetric Designs

The classification of symmetric matroid designs is essentially contained in the theorem of Dembowski and Wagner [4]. We spell it out for completeness.

Theorem 7. (Dembowski-Wagner) The parameter set (v,k,λ) is matroidal if and only if it is one of the following types:

(a) $\lambda = 1$, that is the parameters correspond to a projective, not necessarily desarguesian plane.

(b) It is the parameter set of a design derived from a projective space PG(n,q). That is for some prime p and positive integer n,

$$v = \frac{(q^{n+1} - 1)}{q - 1}, \quad k = \frac{q^n - 1}{q - 1}, \quad \lambda = \frac{q^{n-1} - 1}{q - 1}$$

where $q = p^r$, for some positive integer r.

Proof That a parameter set satisfying (a) or (b) is matroidal is clear. The converse is most easily seen from the version of the Dembowski-Wagner theorem presented by Kantor [8].

This states that if the blocks of a symmetric design $D(v,k,\lambda)$ satisfy the hyperplane axioms and also the conditions (i)$\lambda > 1$ (ii) for distinct points p,q there is a block containing p but not q and (iii) $v - 2 \geq k > \lambda$; then the design is the set of hyperplanes and points of a finite projective space. Condition (i) is covered by (a). Conditions (ii) and (iii) are degeneracy restrictions.

An easy consequence of Theorem 1 is the following.

Theorem 8. *If (v,k,λ) is matroidal then the complementary set $(\overline{v},\overline{k},\overline{\lambda})$ is non-matroidal.*

Proof The complementary set $(\overline{v},\overline{k},\overline{\lambda})$ satisfies

$$\overline{v} = v, \quad \overline{k} = v - k, \quad \tilde{\lambda} = v - 2k + \lambda .$$

Suppose that (v,k,λ) is of type (a), that is $\lambda = 1$. Then if $\tilde{\lambda} = 1$, $v = 2k$ and the existence conditions

$$k - \lambda = k^2 - \lambda v$$

become

$$k - 1 = k(k - 2)$$

which is impossible.

If $(\tilde{v},\tilde{k},\tilde{\lambda})$ is of type (b) the condition $\lambda = \tilde{v} - 2\tilde{k} + 1$ implies that for some prime power q

$$q^{n+2} - 2q^{n+1} + q = 1$$

which is again not possible unless $q = 1$.

Similarly if (v,k,λ) is matroidal of type (b), elementary calculations show that it is impossible for $(\tilde{v},\tilde{k},\tilde{\lambda})$ to be either of type (a) or type (b).

4. Matroidal BIBDS

The situation with BIBDS seems much more complicated,

and so far I have not been able to find necessary and sufficient conditions for a set of parameters (b,v,r,k,λ) to be matroidal.

<u>Theorem 9</u> <u>The following BIBDS are matroidal:</u>
(i) <u>Any design derived from a Steiner system</u> $S(d,k,n)$.
(ii) <u>Any design whose blocks are the flats of a given dimension of either a projective or affine geometry.</u>

Notice that since designs with $\lambda = 1$ are Steiner triples we have

<u>Corollary</u> <u>Any set of design parameters</u> $(b,v,r,k,1)$ <u>is strongly matroidal</u>.

Designs of type (ii) are well known to satisfy the matroid axioms. A design of type (i) is a d-paving geometry in the terminology of Crapo and Rota [2].

The case $\lambda = 2$ is equally easy to settle.

<u>Theorem 10</u> <u>Any set of design parameters</u> $(b,v,r,k,2)$ <u>is non-matroidal</u>.

<u>Proof</u> Let D be a design with $\lambda = 2$. If x,y, are distinct elements they must belong to exactly 2 blocks, B_1, B_2. Unless $B_1 \cup B_2$ contains every point of D, then D is non-matroidal and simple calculation shows that this can only occur in the degenerate symmetric case $D = D(4,3,2)$.

<u>Theorem 11</u> <u>Necessary conditions for</u> (b,v,r,k,λ) <u>to be matroidal when</u> $\lambda > 2$, <u>are that the following hold</u>:

(i) $\lambda(k - 2) + 2 \geq v$

(ii) $b\binom{k}{3} \geq \binom{v}{3}$

Proof Since $\lambda > 2$, if x,y,z, are distinct points $\{x,y\}$ is contained in two distinct blocks and hence $\{x,y,z\}$ must be contained in at least one block. This proves (ii). Suppose that $\{x,y\} \subseteq B_1 \cap B_2$. Then the λ blocks containing $\{x,y\}$ must cover the points of the design. This proves (i).

Theorem 12 A necessary condition that the parameters (b,v,r,k,λ), $(\lambda>2)$ are matroidal is that for some i $(2 \leq i \leq k-2)$ the following conditions are satisfied.

(i) $b\binom{k}{i+1} \leq \binom{v}{i+1}$

(ii) $(v-i)/(k-i)$ is an integer p not greater than λ.

(iii) $k \leq 1 + p(i-1)$

Although these conditions seem complicated they are surprisingly easy to check (at least for small values of k and λ).

Proof Call a design D, u-starshaped if there exists a set X and blocks $B_1, \ldots, B_p$ such that

$$|X| = u,$$

$$B_i \cap B_j = X, \qquad (i \neq j;\ 1 \leq i, j \leq p),$$

$$\bigcup_1^p B_i = S.$$

Let $u = \max_{i \neq j} |B_i \cap B_j|$, where the maximum is taken over all blocks of D. Suppose that $X = B_1 \cap B_2$ and $|B_1 \cap B_2| = u$. Then for any $y \notin B_1 \cup B_2$, $X \cup \{y\}$ must be contained in some block, say B_3 and

$$B_3 \cap B_1 = B_3 \cap B_2 = X$$

for otherwise u is not a maximum. Hence continuing in

this way we eventually exhaust the points of D, and notice that every matroidal design is u-starshaped. Also since $\lambda > 2$, $u \geqslant 2$ and therefore $p \leqslant \lambda$. In order that $D(v,b,r,k,\lambda)$ be u-starshaped it is clear that $(v-u)/(k-u)$ must be an integer $p \leqslant \lambda$. Moreover any $(u+1)$-subset of points of D must belong to at most one block of D. A necessary condition for this is that

$$\binom{v}{u+1} \geqslant b\binom{k}{u+1} \quad .$$

To obtain (iii) notice that in a non-degenerate design any element $x \in X$ belongs to more than λ blocks. Clearly if B contains x, $B \notin \{B_1,\ldots,B_p\}$ then $B \cap B_i \leqslant u$ and hence

$$k \leqslant 1 + p(u-1).$$

To see that this is a stronger result than Theorem 11 consider the case $\lambda = 3$. Theorem 11 gives that a necessary condition a parameter set $(b,v,r,k,3)$ be matroidal is that $v \leqslant 3k-4$. Using Theorem 12 and non-degeneracy conditions we see that

$$k \leqslant 1 + 3(u-1)$$

$$3(k-u) = v - u$$

which gives

<u>Corollary</u> <u>A necessary condition that</u> $(b,v,r,k,3)$ <u>be matroidal is that</u>

$$3v \leqslant 7k - 4 \ .$$

Finally we obtain partial analogues of Theorem 2.

<u>Theorem 13</u> <u>Any design with</u> $\lambda = 1$ <u>has a non-matroidal complement and any design with</u> $\lambda = 2$ <u>has a strongly matroidal complement</u>.

Proof Remembering that the set complement of a hyperplane of a matroid is a circuit of the dual matroid it is routine to check that apart from degenerate cases, when $\lambda = 1$ the blocks of a design do not satisfy the circuit axioms while if $\lambda = 2$ the blocks do satisfy the circuit axioms.

5. Conclusion

Using the above results it is usually easy to decide whether or not a 'low-parameter' design is matroidal. For example they classify all but five of the parameter sets given in Table 1 of Hall [5]. The parameter sets still undecided have Hall numbers (49,83,93,97,98).

Finally I mention some questions to which I would like to know the answer but not having spent much time on them do not feel justified in calling them 'unsolved problems'.

1. Murty [9] in his work on equicardinal matroids calls a matroid proper if it is connected and for any two distinct elements p,q, there exists a hyperplane containing p but not q. He goes on to prove that every proper, equicardinal, binary matroid is either a uniform matroid (trivial design) or a design derived from a projective or affine geometry over GF(2). This does not hold for non-binary matroids. The class of identically self dual matroids on a set of 4n elements derived in [1] are proper, equicardinal, but are not the blocks of a design. However I do not know of any proper equicardinal matroid in which each element does not belong to the same number of hyperplanes. If this is true then all equicardinal matroids can be derived from partially balanced incomplete block designs.

2. Does there exist a matroid design not derivable from a projective or affine geometry or a Steiner system?

3. If $\underline{M}$ is a non-trivial, proper equicardinal matroid such that for any two hyperplanes H_i, H_j, $|H_i \cap H_j| = \lambda > 1$ then is $\underline{M}$ a design matroid?

Acknowledgement. I would like to acknowledge many helpful conversations with J.A. Bondy and U.S.R. Murty. I am also indebted to Dr Murty for showing me Kantor's paper. I am also very grateful to Dr. D.R. Woodall for some very helpful criticisms.

References

[1] Bondy, J.A., and Welsh, D.J.A., Some results on transversal matroids and constructions for identically self dual matroids, Quart. J. Math. (Oxford) (to appear).

[2] Crapo, H.H., and Rota, G.C., Combinatorial Geometries, M.I.T. Press (to appear).

[3] de Sousa, J., and Welsh, D.J.A., A characterisation of binary transversal matroids, J.Math.Anal. and Appl. (to appear).

[4] Dembowski, P., and Wagner, A., Some characterisations of finite projective spaces, Arch.Math., 11, (1960), 465-469.

[5] Hall, M., Combinatorial Theory, Blaisdell, Waltham, (1967).

[6] Harary, F., Piff, M.J., and Welsh, D.J.A., On the automorphism group of a matroid, Journ.Discrete Math., (to appear).

[7] Harary, F., and Welsh, D.J.A., Matroids versus graphs: The Many Facets of Graph Theory, (Springer Lecture Notes), 110, 155-170.

[8] Kantor, W.M., Characterisations of finite projective and affine spaces, Canad.J.Math., 21, (1969), 64-75.

[9] Murty, U.S.R., Equicardinal matroids, J.Comb.Theory, (to appear).

[10] Murty, U.S.R., Equicardinal matroids and finite geometries, Combinatorial Structures and their Applications, (Gordon and Breach), (1970), 289-293.

[11] Piff, M.J., Further results on the automorphism group of a matroid, (to appear).

LISTE DES PARTICIPANTS

		Nationalité
BERMOND, J.C.	Centre de Mathématiques Sociales, 17, rue Richer, PARIS 9è	F
BRUALDI, R.A.	Department of Pure Mathematics, The University, SHEFFIELD	G.B.
BRUTER, C.P.	Département de Mathématiques, Faculté des Sciences, 29-N BREST	F
CHEIN, M.	Département de Mathématiques Appliquées CEDEX 53, 38-GRENOBLE	F
DINOLT, G.W.	Department of Pure Mathematics, The University, SHEFFIELD	G.B.
DUCHAMP, A.	Département de Mathématiques, Faculté des Sciences, 14 - CAEN	F
FOURNIER, J.C.	Institut H. Poincaré, 11, rue P. et M. Curie, PARIS 5è	F
HOCQUENGHEM, S.	Conservatoire National des Arts et Métiers, 292, bd St. Martin, PARIS 3è	F
INGLETON, A.W.	The Mathematical Institute, The University, OXFORD	G.B.
JOLIVET, J.L.	Département de Mathématiques, Faculté des Sciences, 14 - CAEN	F
LAS VERGNAS, M.	Institut H. Poincaré, 11, rue P. et M. Curie, PARIS 5è	F
LECLERC, B.	Centre de Mathématiques Sociales, 17, rue Richer, PARIS 9è	F
MASON, J.H.	Department of Mathematics, Open University, BLETCHLEY (Buchs)	G.B.
MONIEZ, F.	Institut de Recherche des Transports B.P. 28, 94 - ARCUEIL	F
MONJARDET, B.	Centre de Mathématiques Sociales, 17, rue Richer, PARIS 9è	F

PERFECT, H.	Department of Pure Mathematics, The University, SHEFFIELD	G.B.
ROSENSTHIEL, P.	Centre de Mathématiques Sociales, 17, rue Richer, PARIS 9è	F
SAKAROVITCH, M.	Institut de Recherche des Transports B.P. 28, 94 - ARCUEIL	F
SOUSA de, J.	The Mathematical Institute, The University, OXFORD	G.B.
TUTTE, W.T.	Department of Combinatorics and Optimization, U. of Waterloo, WATERLOO	C
VAMOS, P.	Department of Pure Mathematics, The University, SHEFFIELD	G.B.
WELSH, D.J.A.	The Mathematical Institute, The University, OXFORD	G.B.
WILSON, R.J.	The Mathematical Institute, The University, OXFORD	G.B.